Des

A handbook for Integrate

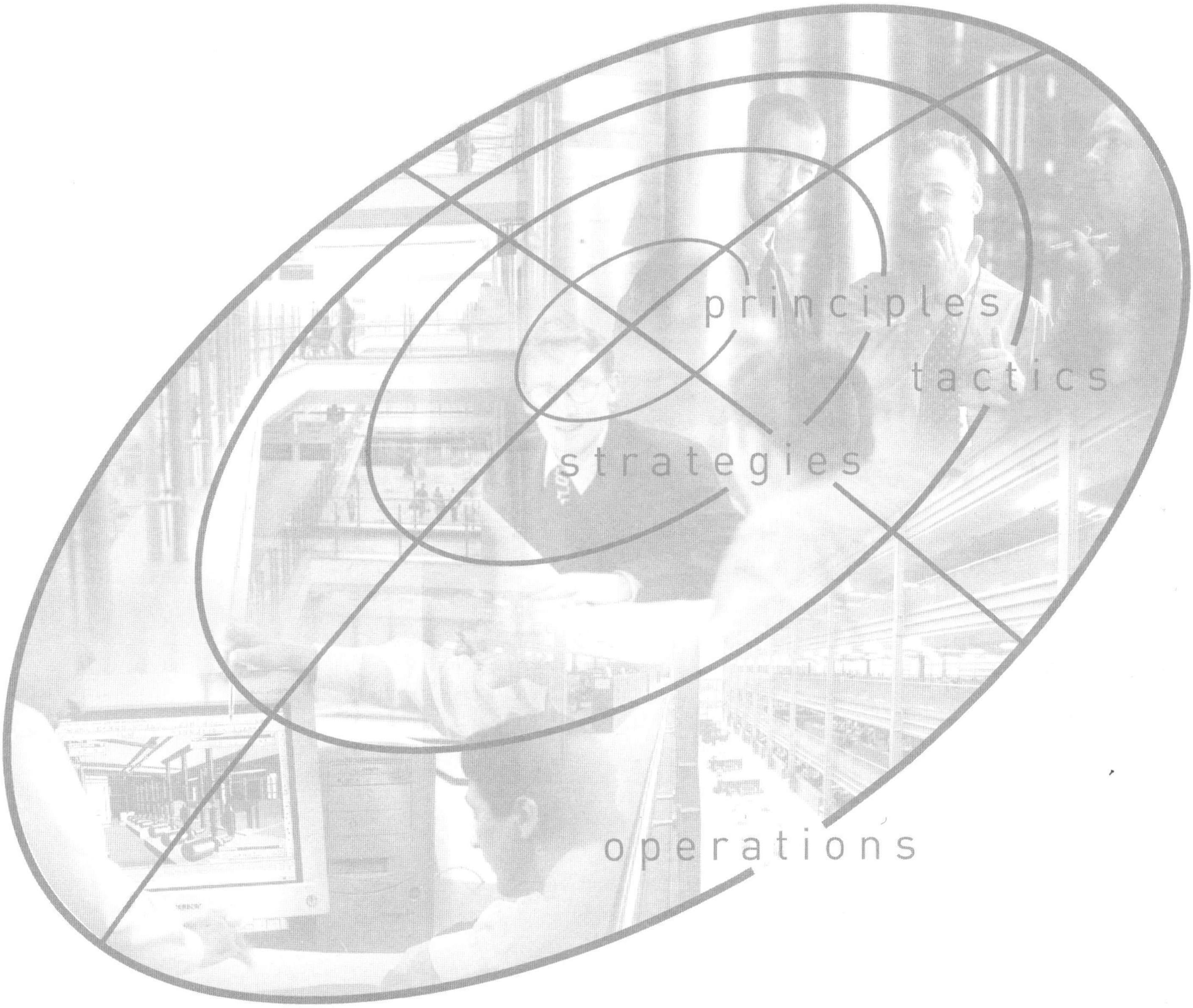

principles

tactics

strategies

operations

S. Austin, A. Baldwin, J. Hammond, M. Murray, D. Root, D. Thomson and A. Thorpe
of Loughborough University

in partnership with AMEC, DTI and EPSRC

Loughborough University
Loughborough
Leics.
LE11 3TU
Telephone +44 (0) 1509 222608
www.lboro.ac.uk

AMEC
Timothy's Bridge Road
Stratford-upon-Avon
Warwickshire
CV37 9NJ
Telephone +44 (0) 1789 204288
www.amec.com

Further copies of this handbook are available from:

The Customer Services Department
Thomas Telford Limited, Units I/K
Paddock Wood Distribution Centre
Paddock Wood, Tonbridge
Kent TN12 6UU
Telephone: 020 7665 2464
Fax: 020 7665 2245

www.thomastelford.com
First published 2001

ISBN 0 7277 3039 8

© Loughborough University, 2001

The conclusions and recommendations presented here are those of the researchers and not necessarily those of Loughborough University and AMEC.

All rights, including translation, reserved. Except as permitted by the Copyright, Designs and patents Act 1988, no part of this publication may be reproduced, stored in a retrieval system or transmitted in any form or by any means, electronic, mechanical, photocopying or otherwise, without the prior permission of the Publishing Director, Thomas Telford Publishing, Thomas Telford Limited, 1 Heron Quay, London E14 4JD.

This book is published on the understanding that the authors are solely responsible for the statements made and opinions expressed in it and that its publication does not necessarily imply that such statements and/or opinions are or reflect the views or opinions of the publishers. While every effort has been made to ensure that the statements made and the opinions expressed in this publication provide a safe and accurate guide, no liability or responsibility can be accepted in this respect by the authors or publishers.

Design by ACP Graphics, Stratford-upon-Avon, UK
Printed in Great Britain by Hobbs the Printers, Totton, Hampshire

Foreword

Design Chains

I am pleased to welcome this new handbook "*Design Chains*".

The construction industry still faces great pressures to innovate in order to improve its performance and make a wider contribution to the economy.

In my 1994 report on the UK construction industry "*Constructing the Team*" I highlighted the fuzzy edge between designers and the supply chain as a significant cause of poor performance in construction. This was, and unfortunately remains, a key issue which the industry needs to address.

Design Chains is an excellent tool with which to bridge the gap. It describes an initiative which should do much to harness more closely together design creativity and project delivery, and address the frequently imperfect link between designers and the supply chain.

Design Chains is about why big and small companies need each other, why their relationships should be long term and how they can sustain a relationship even when they are not always involved in the same projects. It is a manual about how to form networks and share the latest innovative thinking.

The underpinning research behind this handbook has been conducted as a collaborative project between supply chain partners from industry and one of our leading Universities. This in itself is a major achievement in an industry which has a low level of investment in Research and Development.

Design Chains represents innovative thinking for the companies who have developed the concept to give them critical advantage in the market. The fact that they are sharing it here is a welcome sign that this industry is accepting the spirit and reality of real team building.

This is an important initiative. It deserves to be read.

I commend it to the whole construction team, led by clients and throughout the supply chain.

Sir Michael Latham

Acknowledgements

Department of Trade and Industry
Engineering and Physical Sciences Research Council
Loughborough University

Clive Balsom	Crown House Engineering
Glenn Ballard	University of California, Berkeley
David Binks	Hathaway Roofing
Mark Boorman	AMEC
Colin Bunce	AMEC Graphics
Chris Carter	Loughborough University
James Choo	University of California, Berkeley
Ian Coverdale	Hathaway Roofing
Colin Davidson	AMEC
Department of Civil and Building Engineering	Loughborough University
John Duggan	AMEC
David Earp	Colt International
Keith Edgell	AMEC
Scott Fernie	Reading University
John Fisher	AMEC
Tim Fry	AMEC
Adam Greene	Loughborough University
Kevin Hall	Hathaway Roofing
Robert Jackson	DTI
Keith Johnson	E-Squared
Carl Leatham	MSS
Baizhan Li	Reading University
Ron Linn	Galloway Group (Northern)
David Lowe	Senior Hargreaves
Malcolm Mason	Environmental Air Contracts
Terry McCarthy	DTI
Martin Murphy	AMEC
Caroline Neale	Loughborough University
Andrew Newton	AMEC
Andy Page	AMEC Graphics
Brian Pimlott	AMEC
Adam Poole	Adam Poole Associates
Matt Quinn	AMEC
Peter Shields	Briggs Roofing
Lawrence Simpson	Hilton Building Services
John Steele	AMEC
Paul Stubley	MSS Clean Technology
Teresa Sykes	AMEC
Bill Taplin	AMEC
Colin Taylor	E-Squared
Iris Tommelein	University of California, Berkeley
Robert Van Zyl	AMEC
Owen Vickery	AMEC
Paul Waskett	AMEC
Neil Wilson	Honeywell
Jacqui Williams	EPSRC
Craig Woodall	AMEC

Contents

About this Handbook

Origins

This handbook presents the work of the Integrated Collaborative Design (ICD) research project, a combined industry and academic initiative between Loughborough University and twelve construction companies.[1] Financial support was given by the EPSRC and the DTI as part of the LINK Integration in Design and Construction programme.[2]

This handbook presents new thinking in supply chain management derived from this research, how this impacts on the flow of design information within projects,[3] and on the wider effects this has on the strategic relationships (business domain) between companies.[4] ICD places the design processes and the management of design information at the centre of project management practice.

Intended Readership

This handbook has been written to be of value to a wide readership, from clients to specialised suppliers and should be particularly useful if you are concerned with:

- providing design information for use by others;
- receiving and applying design information from others;
- creating suitable frameworks to ensure design information can be used effectively in both business domain and project domain activities;
- applying design information effectively in projects.

Handbook Structure

Part 1: Introduction Explains the design chain concept and introduces the need for ICD.

Part 2: Integrated Collaborative Design Describes the principles and practices of ICD.

Part 3: Applying ICD Principles Gives a detailed discussion of how an organisation can adopt and apply the three principles of ICD.

Part 4: Applying ICD Practices Guides you through the selection of relevant ICD practices, depending on your role in your organisation and the role you typically perform in a design chain (as either a provider or receiver).

Part 5: The ICD Practices Describes ICD practices that can be used to enhance your ability, and your organisation's ability, to perform within a design chain.

---

[1] The 12 construction companies are: AMEC; Briggs Roofing and Cladding; Colt International; Crown House Engineering; Environmental Air Contracts; E-Squared; Galloway Group (Northern); Hathaway Roofing; Hilton Building Services; Honeywell; MSS Clean Technology; Senior Hargreaves.

[2] EPSRC: Engineering and Physical Sciences Research Council (www.epsrc.ac.uk); DTI: Department of Trade and Industry (www.dti.gov.uk). Project reference: LINK IDAC 435, grant GR/M11240.

[3] Elements of this work build on the Analytical Design Planning Technique (ADePT), winner in the Achievement through Innovation and Supreme categories at the 1999 Quality in Construction Awards. See www.adeptmanagement.com for more information.

[4] ICD makes an important distinction between company activities in the project and the business domains. See definition of key terms at the end of this section or glossary for details.

Further Information

Associated with this handbook are two websites where supporting material and documentation can be found.[5]

Key Terms

This handbook introduces the design chain as a core new concept. Details are given in the glossary, but for now, the key terms to keep in mind are as follows.

Principles underpin an organisation's approach to ICD to manage its collaboration with others when devising design solutions. Each principle must be adopted by an organisation to facilitate the use of ICD practices.

Practices are used to apply the principles of ICD in the everyday activities of an organisation. A practice may be a:

> **Strategic Practice**, which will help an organisation plan elements of its ICD approach; a
>
> **Tactical Practice**, which will help an organisation respond to circumstance as it implements its ICD strategy - predominantly by making a variety of tools and techniques available for use; or an
>
> **Operational Practice**, which will help an organisation put its ICD approach into action in everyday business and project activities.

Domains define the focus of an organisation's approach to ICD. The domain considered may be either the:

> **Business Domain**, in which long-term relationships between organisations allow mutually beneficial strategies to be developed and cultural alignment achieved; or the
>
> **Project Domain**, in which organisations collaborate to achieve a common objective that they could not complete if working in isolation.

Roles define how an individual within a design chain processes information. Individuals may be either a:

> **Provider**, who predominantly generates design solutions for use by receivers; or a
>
> **Receiver**, who generally uses design solutions generated by a provider.

These terms appear throughout this handbook and, when encountered, should be read with the above definitions in mind. You may also find it useful to refer to the Essentials of ICD, on page 57.

The ICD terminology is explained in detail in the Glossary at the back of this handbook.

---

[5] For more information about design chains, visit www.designchains.com. For more information about the research that gave rise to this handbook, visit www.abouticd.com.

Part 1: Introduction

1.1 Design Chains and the Need for an Integrated Approach

Why this, why now?

The construction industry is adept at delivering highly complex products. It differs from much of engineering in that its products are often unique to the specific needs of its clients and are generally assembled in unfamiliar locations. Construction projects involve relationships between many organisations and thousands of processes. Over the years, the industry has evolved highly developed methods of accommodating this complexity and has embedded them in the management of projects. Organisation relationships are controlled through well-established contractual forms and the delivery and assembly of components to dispersed locations is being continually refined with ever-improving working methods and assembly technologies.

> *The ICD terminology is explained in detail in the Glossary at the back of this handbook.*

However, design - that part of construction that needs to be in place before the physical work can begin - remains an area where the complexity of the process is not yet understood and managed to the same standard. For much of construction, design exists as an unknown process - a 'black box'. It occurs at a particular time in the schedule, is performed by different parties, is creative and is often poorly controlled. This handbook is about understanding the process more fully and the roles individuals play in it.

> *Design chain*
> *A design chain is created when organisations collaborate in the development of project design information.*

Clients are demanding that the industry improves how it delivers projects and that it should learn from best practice, not only within the industry, but also from other industries. One solution is to build what we have termed *design chains*. Design chains provide the means by which companies can understand and work with each other and can develop sustaining relationships that address and accommodate the full complexity of the design process. These are formed when organisations collaborate in the development of project design information.

What is a Design Chain?

A design chain is that part of the project supply chain function which is focused on design. A design chain is created when organisations, which repeatedly collaborate in project supply chains, co-ordinate their roles to develop project design information for their mutual benefit. To achieve this, organisations make a series of provisions in both their long-term *business domain* relationships and their short-term *project domain* relationships. These provisions are made in the light of each organisation's understanding of the design information requirements of others. One organisation will plan the production of its design information to help others to use it, and vice versa. In the same way that supply chains have revolutionised the production of goods and services, design, which exists as separate processes carried out by different organisations, can be managed as a 'chain' across organisational boundaries.

In summary, design chains:

- allocate design processes on the basis of an organisation's technical competency;
- ensure that the organisations belonging to design chains have the ability to work together; and
- align organisations to improve their combined effectiveness and to ensure that processes are not duplicated nor missing in the chain.

To understand the role of an organisation within a design chain, it is important to distinguish between design and design information. Design, naturally enough, is the act of designing. It is performed by each organisation on the design information that passes between them. Design information is the raw material on which individual designers act. It is comprised of problems and solutions, and remains inert until it is either applied or amended by a designer (through the act of design).

> **(i)** *Design chains are an aspect of project supply chains that is managed and co-ordinated to benefit their member organisations and project clients through improved co-ordination of design processes.*

Design Chains and Supply Chains

Design is iterative, making the exchange of design information quite different from that of physical goods in the traditional product-based supply chain.

In a product-based supply chain, physical goods move along the chain with each tier adding value through some manufacturing process until the completed product (a car for example) is available to the end-user (Figure 1.1). The product becomes increasingly complex as it passes through each stage (e.g. by combining standard components), as it is transformed from raw materials into a customised product.

At each stage along the supply chain, demand is communicated through a flow of information down the chain as an order or contract placed. Demand moves in the opposite direction to the physical product, which flows up the chain.

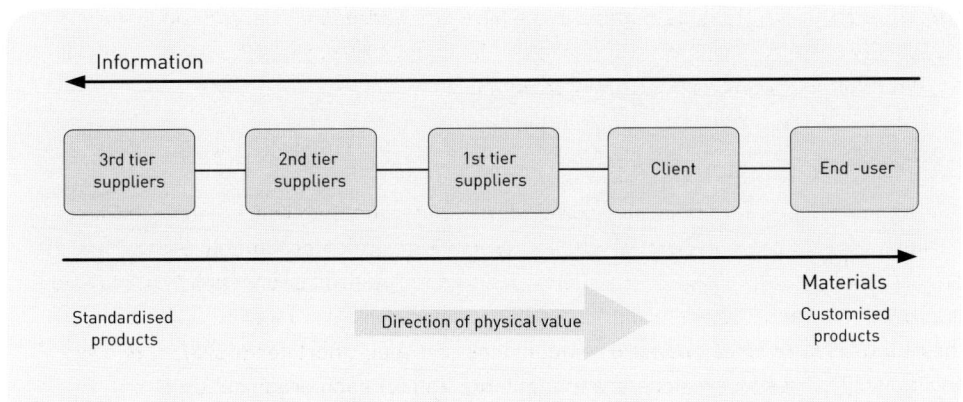

Figure 1.1:
A simplified model of a supply chain supplying a physical product

In the case of a construction project, a similar supply chain can be envisaged, where value is added when construction products flow up the chain. Standardised products (such as glass, steel and cement) are combined to provide more specialised components and systems (cladding and air conditioning, for example). The suppliers in the chain will be made up of a combination of design organisations (consultants); contractors (including those with design and/or management specialisms); and sub-contractors, depending on the method of procurement.

As with the supply chain, where standard products are converted into bespoke facilities, solutions in the design chain become increasingly specialised and complex. Standard solutions (such as standard design details) are combined to provide more comprehensive and bespoke design solutions.

Design chains are concerned with the flow of design information between the organisations collaborating on a project. It differs from a supply chain in not having information flowing in one direction and material in the other. In design chains, information flows both ways (Figure 1.2). However, the ICD design chain distinguishes between two types of information: problems (which frame requirements) and solutions (which provide a solution to a previously defined problem). For example, a performance specification may be compiled to frame the design problem for a lift installation. The specification[6] defines the problem and the resulting lift design from the manufacturer is a solution to that problem. By understanding the relationship of design problems to design solutions, individuals can define themselves as either a *provider* or a *receiver* (or both) of design information for a particular level (tier) in the design chain. As projects progress, individuals may switch between roles. For example, a quantity surveyor may provide budget costings during pre-contract work and may receive applications for payment during construction.

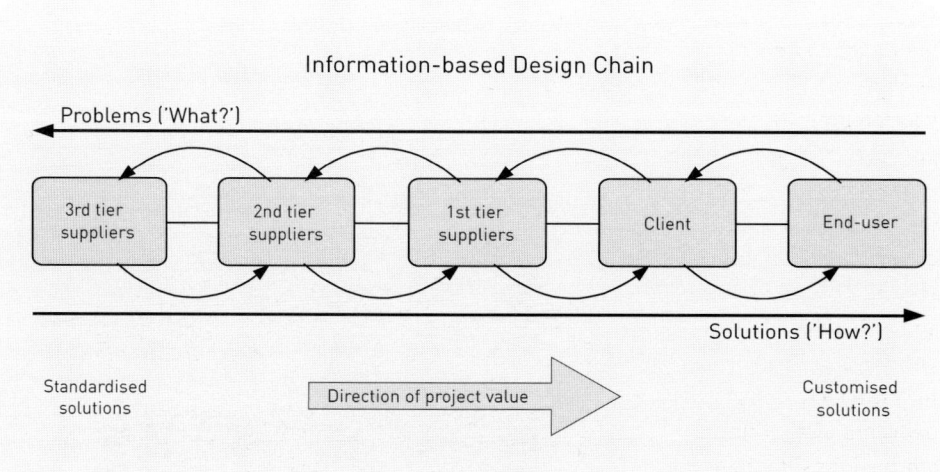

Figure 1.2:
A simplified model of
a design chain providing
a completed design

---

[6] At another interface or tier within the design process, the specification is itself a solution to a design problem.

By understanding the flow of problems and solutions in a design chain the roles of individuals as providers and receivers can be defined (Figure 1.3). A receiver initiates the exchange of design information by setting a problem (e.g. by stating requirements). A corresponding provider responds by generating a design solution and communicating it to the receiver thereby completing the cycle of design information flow.

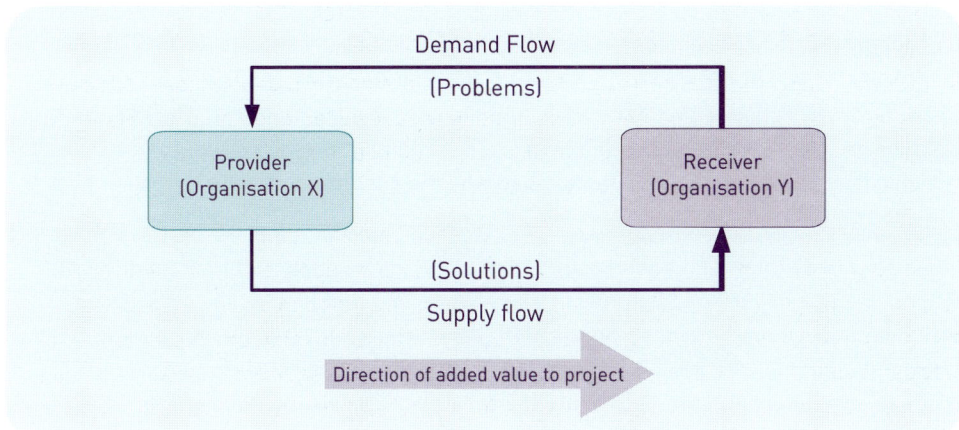

Figure 1.3:
The cycle of design information flow between receiver and provider across organisation boundaries

What is Integrated Collaborative Design?

ICD is an approach that establishes design as the common thread linking organisations together. At its core are a set of **principles**, which are applied to business domain and project domain activities and are supported by strategic, tactical and operational **practices**. It provides a basis for managing relationships between strategic partners, for building design chains and distinguishing between provider and receiver roles within the design chain.

ⓘ *Integrated Collaborative Design is an approach that establishes design as the common thread linking organisations together.*

This handbook has been written to describe the ICD approach from four perspectives: business domain; project domain; provider and receiver. As can be seen in Figure 1.4, it is useful to see these four perspectives as quadrants of a circle, with business domain and project domain perspectives existing on one axis and the provider and receiver perspectives on the other.

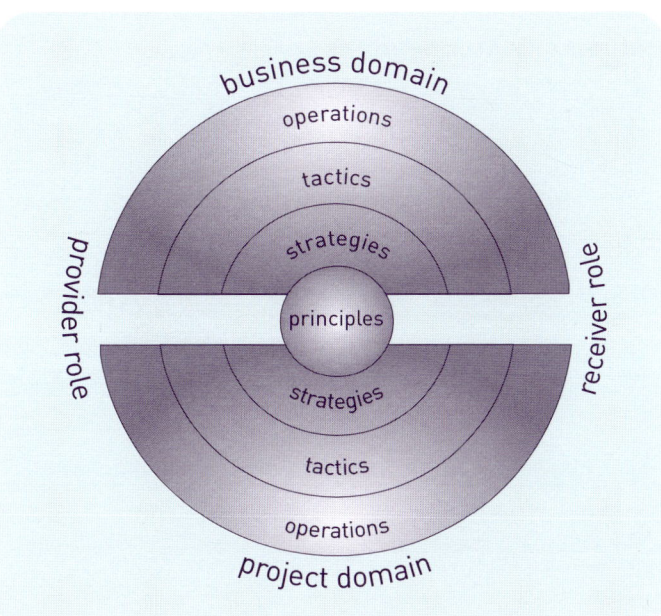

Figure 1.4:
The basic elements of an ICD approach to design management

1.2 Setting the Scene

Construction in the 21st Century

There is a widely held belief that construction is changing. Recent reports by Latham in 1994[7] and Egan in 1998[8] have highlighted the need for this change and have begun to create a climate for it to take place. Helping to drive this change is the use of information technology (IT). It has created new forms of industrial organisation (such as virtual companies) in which design responsibilities are distributed across the supply chain, following the practice of the automotive and the aerospace industries.[9] The dramatic productivity improvements achieved by these industries in the last 30 years have raised the question, particularly with clients, as to why the construction industry has not yet achieved similar gains.

Books such as The Machine that Changed the World [9] have shown the dramatic effect that supply chain management can have on quality and productivity.

Following the Latham and Egan reports, and with clients actively encouraging more innovative ways of working, the industry has begun to adopt new working practices, including supply chain management (SCM), benchmarking, lean production and total quality management (TQM).[10] However, the improvement of design management methods has been hindered by the intuitive and iterative nature of design.
This makes it difficult to model, plan, and manage design in the same way as more sequential processes.[11] Management techniques are now becoming sophisticated enough to be applied to the previously poorly understood process of design and, as a result of process-mapping research,[12] it is becoming increasingly easier to plan for design decision-making with the same rigour as other parts of the process.

These tools pave the way for the use of design chains to manage design processes between organisations and to optimise the total design solution in the same ways that supply chains have revolutionised the production of goods and services. An ICD approach provides organisations with design management methods and a framework for collaborative working. These methods will assist organisations wanting to integrate their ways of working and will, in turn, allow them to achieve more together, thus meeting the demands of clients (who view construction as any other complex process, which requires efficient and effective management).

---

[7] Latham, M. (1994) *Constructing the Team: Joint Review of Procurement and Contractual Arrangements in the United Kingdom Construction Industry*, HMSO, London.

[8] Egan, J. (1998) *Rethinking Construction: The Report of the Construction Task Force to the Deputy Prime Minister, John Prescott, On the scope for improving the quality and efficiency of UK construction*, Department of the Environment, Transport and the Regions, London.

[9] Womack, J.P., Jones, D.T., Roos, D. (1990) *The Machine That Changed The World*, Macmillan, New York.

[10] A number of groups and initiatives have been established to address issues identified by the Latham and Egan reports, including: the Construction Clients Forum (CCF); the Movement for Innovation (M4I); and the Construction Best Practice Programme (CBPP). All have promoted new working practices to some degree.

[11] See Austin et al. (2000) Integrating Design in the Project Process, *Proceedings of ICE Civil Engineering*, Volume 138, pp. 172-182, for more information on ADePT.

[12] The research was conducted by the team that developed ADePT. ICD is derived in part from ADePT, which makes extensive use of IT to enable a new approach to design management. See www.adeptmanagement.com for more information.

Increasing Design Complexity and Fragmentation

When Wren designed St Paul's Cathedral, he acted as both engineer and architect. Since his day, the design and construction process has become progressively more fragmented due to the growth in specialisation and complexity of construction methods and technologies. With this specialisation, there has been a corresponding increase in the number of organisations and people with design responsibility on a project. In Wren's day, the processes and iterative nature of design were hidden within a single mind. The task today is to understand and to manage these processes across a team of designers.

Integrating Activities across Organisations

A big part of the challenge in running a team of designers from different companies is in managing the relationships between them. Organisational interfaces have always been problematic in construction and are often a source of dispute. Integrated collaborative design must overcome this.

The tradition of single design disciplines in the construction industry has led to a situation where small firms, with limited skill bases, have to contend with a great variety of contractual arrangements and inter-firm relationships. Even larger multidisciplinary firms that might have structural, building services and architecture expertise under one roof, still face problems of interdisciplinary communication.

Theoretically, multidisciplinary firms or teams should be able to provide integrated design. But, even when inter-organisational boundaries have been removed, the existence of different design disciplines may give rise to 'silo-culture', where different departments develop their own values and objectives, which may differ from those of the company. Just as design management can bring together the different creative cultures that exist between organisations, so can it be used within organisations to optimise the flow of information between functional groups and design disciplines.

Historically, the industry sees construction contracts as the means of controlling relationships between project participants. This is often regardless of whether design is actually covered under the contract or whether it is generated separately through design commissions; procurement routes, such as design and build (D&B) and design manage construct (DMC), have been attempts to secure contractual integration.

> *Collaborative design requires an easy flow of information between all participants.*

Standard forms and conditions of contract have tended to focus on the possibility of failure, by defining the means and channels of communicating design (and other) information between the project parties and avenues of recourse. In trying to manage the design processes within these rigid structures, the contractual route begins to break down. Collaborative design requires an easy flow of information between all participants.

Contractual arrangements that seek to define and control the flow of design information hinder design. This is because, as they tend to be concerned with design liability and risk allocation, they act to ensure minimum standards and do not encourage creative leaps forward. The issues surrounding collaborative design (such as the need for flexibility) are generally too complex to be reflected accurately in standard contracts. Figures 1.5 and 1.6 illustrate this by comparing the complexity of exchange when transferring physical goods with the exchange of design information.

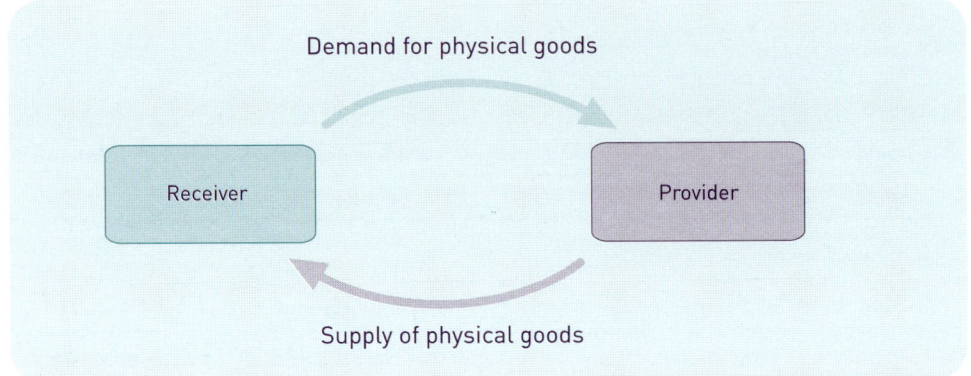

Figure 1.5:
The simple exchange
of physical goods

Figure 1.5 shows a simple, single-stage commodity exchange (e.g. bricks) between a provider and a receiver. Figure 1.6, in contrast, shows the more complex exchange that takes place when a design solution is developed through an iterative process.

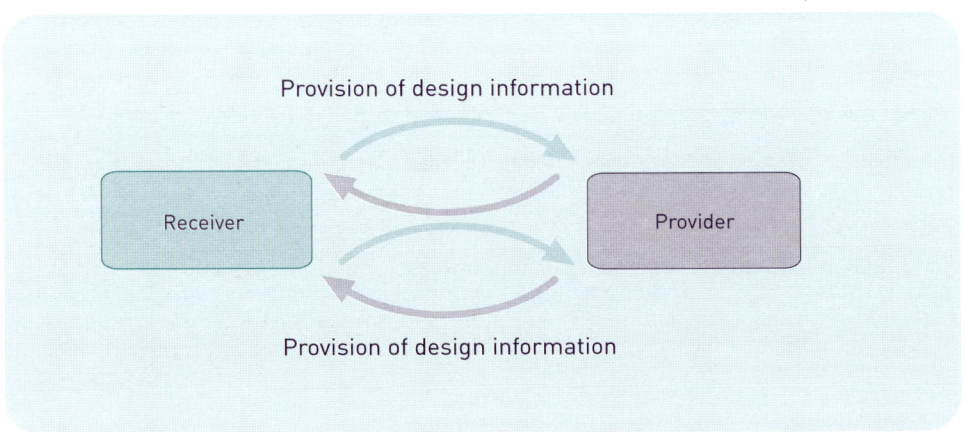

Figure 1.6:
The complex exchange
of design information

Clearly, the multiple flows of information between organisations undertaking collaborative design mean that the mechanism of controlling the process is more complex. Consequently, the tools and practices used to manage them will be different from those used on more linear processes, such as manufacturing.

The fact that suppliers can also provide design expertise, which needs to be included in the exchange, adds another layer of complexity. It makes sense for suppliers to have this capability because they tend to have an intimate and tacit knowledge of the manufacturing and installation of the products they supply.

What is a Construction Project?

The term 'project' has a specialised meaning for the construction industry. It encompasses the work undertaken to achieve specific goals, but it also describes the relationships between people and/or groups involved in their delivery. In this respect, the project is often described in terms of an organisation in that it has both structure (relationships) and purpose (activities). However, there is a general recognition that construction projects differ from other organisations, such as companies or government agencies, and that these differences can be summarised in three ways:

- **Projects are unique, one-time-only efforts with distinct objectives. If they merely repeat what has been done previously, then they could be managed as ongoing processes rather than as distinct projects.**[13]
- **Resources are used on each project in a unique configuration. Consequently, each project is not only distinct from other ongoing operations, but it is also different from other projects.**
- **The environment in which a project occurs is always different and constantly changing. Project circumstances change, such as legal jurisdictions and market conditions, as do appropriate management mechanisms.**

Consequently, projects are often referred to as 'temporary multi-organisations' or 'temporary coalitions' and, in the construction industry, are considered to be unique.

It is only firms and individuals operating at the margins of the industry - where they might treat the industry, or construction companies, as customers - who might view projects as ongoing operations. A large retailer who is increasing the number of stores, for example, would tend to focus on the similarities between stores and, therefore,would see the building programme as a single, ongoing operation. This contrasts with the mainstream industry view, which would see the construction of each store within the building programme as an individual project because each would be (slightly) different. Similarly, a component manufacturer, selling to construction and other industries, might view its ongoing business as fulfilling a continuous stream of orders, rather than as contributing to individual construction projects.

---

[13] In particular circumstances, such as the repeated delivery of facilities for fast food chains, multiple 'projects' may have fundamental similarities that override any perceived uniqueness. Even with these projects there are always site and infrastructure differences.

If an organisation views the delivery of construction projects as an ongoing continuous activity, then the adoption of a process perspective as a basis for management is often easier. However, with a 'retailer perspective', the focus is on the similarities between projects rather than the differences. The three characteristics, which, for much of the construction industry, differentiate projects are no longer so pronounced:

1 the same components and systems are required for a project;

2 variation in the resources involved turns out to be quite limited due to the role of personal relationships, technical, and geographical constraints; and

3 many of the environmental factors do not impact directly on designers or other construction staff but are 'filtered out' by their organisations' business processes.

In addition to helping organisations deliver individual projects, ICD also helps organisations to manage the business of delivering a stream of projects. It does this by focusing on the similarities between projects and by realising economies of scale: these bring the benefits of coherence, standardisation, production efficiency and cross-functional learning for both the company and the customer.

1.3 Design Management Needs and the ICD Solution

Four key areas emerge that need to be addressed when integrating design across several organisations. They are to:

1 identify individual design tasks and the relationships between them;

2 allocate responsibility for completing design tasks to organisations on the basis of who is best placed to undertake them;

3 manage the smooth exchange of the design information between collaborative partners; and

4 create suitable working environments that aid the delivery process, such as networks of compatible organisations with shared values, cultures and ways of working.

ICD provides a framework and a set of tools and techniques to assist in the management of these key areas. Part 2 of this handbook describes the three ICD principles and the supporting practices that together make up ICD, as shown in Figure 1.4. As will be seen, the practices are applied at a strategic, tactical or operational level in either the business domain or the project domain.

Part 2: Integrated Collaborative Design

2.1 The Principles of ICD

Background

The ICD approach recognises that an important barrier to the successful implementation of change is the tendency to retreat into existing ways of thinking, such as the idea that all projects are prototypes and unique. ICD looks at construction projects in terms of their similarities and seeks to build upon them; for example, by seeing a building as a unique assembly of standard components within technical systems. ICD allows companies to repeat the processes they use to manage these systems and techniques, and also to learn and to reapply the lessons they learn from working collaboratively.

> The ICD terminology is explained in the Glossary at the back of this handbook.

The construction industry differs from other areas of engineering, such as the car industry or computing, in that new technology has had less impact on its production processes. Technology has expanded the capability of other industries enormously and is driving change. While automation in car manufacture has revolutionised that industry, the introduction of cutting edge technology into buildings (such as the use of computational fluid dynamics to develop passive ventilation systems) alters the construction method little - although it does make for a more comfortable and efficient product. Construction is still largely a craft-based industry, in stark contrast to the advanced manufacturing and assembly processes of industries such as the microelectronic and aerospace sectors.

As a result, parts of the construction industry have remained relatively unchanged throughout the post-Second World War period, although projects themselves have become more complex and the construction industry has become more internationalised. One consequence is that many attitudes held by both individuals and organisations have lagged behind the innovative working practices that have been pioneered elsewhere. The construction industry has been the poorer for it.

A typical example is the construction industry's classification of different types of organisation. In construction, labels such as 'consultant', 'contractor' and 'supplier' are loaded terms. They carry a weight of historically shaped values and behaviours. Despite the growing expectation for suppliers to undertake design, the industry tends to cling to these stereotypes and continues to use outdated labels that only occasionally describe the roles the organisations undertake.

ICD avoids these pitfalls by providing a generic language that can be applied to all the organisations involved in the industry, irrespective of their function or size. In doing so, it builds upon the Building Down Barriers initiative,[14] which describes all members of the supply chain as 'suppliers' regardless of how their role would be traditionally described.

---

[14] Holti, R., Noicolini, D., Smalley, M. (2000) *The Handbook of Supply Chain Management*, Tavistock Institute, CIRIA, London.

Unlike Building Down Barriers and other partnering initiatives,[15] which presuppose a specific type of contractual relationship, ICD does not restrict itself to a particular procurement route. Instead, the approach establishes three *principles* at the core of an organisation's business that can be applied to all the projects that the organisation undertakes.

Building Down Barriers was a Ministry of Defence initiative that helped to define a contracting approach called Prime Contracting.[15]

The three ICD principles applied to design management are (Figure 2.1):

- **applying process management** - to identify tasks;

- **adopting supply chain management practices** - to allocate roles; and

- **establishing value frameworks** - to focus design solutions and to hone process management.

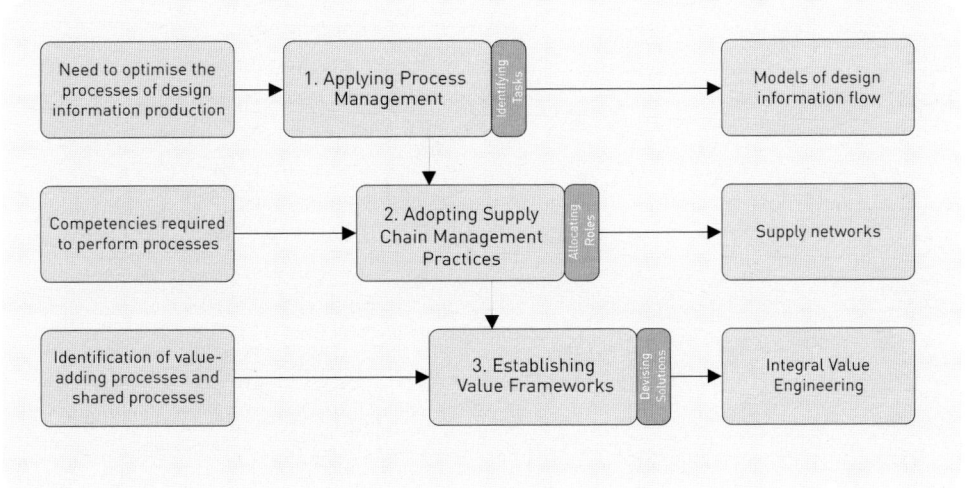

Figure 2.1: The relationship of the ICD principles

By progressively adopting these principles, an organisation's ICD approach evolves, allowing it to benefit from collaboration within *design chains*.

---

[15] Bennett, J., Hayes, S. (1998) *The Seven Pillars of Partnering: A guide to second generation partnering*, Thomas Telford, London.

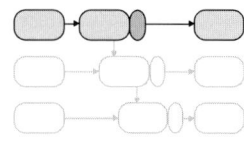

First ICD Principle: Applying Process Management

Within construction, there are a series of activities that repeat between projects. This means that design tasks, and their information requirements, can be captured to produce a 'generic design process model'.

By applying process management, organisations using an ICD approach can represent the complex relationships between them in definitive terms. This increases their understanding of *business domain* and *project domain* activity. By defining their generic design processes, and the corresponding information requirements, organisations can share these definitions with business domain partners allowing everyone to better understand each other's roles and responsibilities.

*Figure 2.2:
Aligning
organisations using
design process
models*

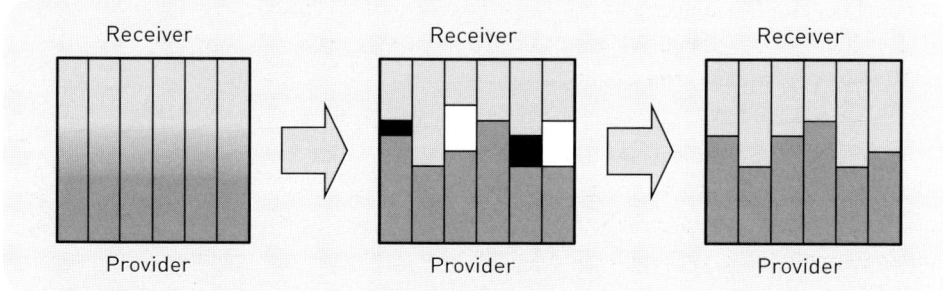

'Greater difficulty can arise over the early involvement of specialist contractors, especially if they are expected to undertake significant detailed design responsibilities of their own. If they are taken on as domestic contractors by a main contractor post-tender, this may affect their ability to work directly in conjunction with the designers' Latham.[16]

Furthermore, process management creates opportunities to optimise supplier involvement and to minimise design overlaps and/or gaps in the scope of project works (Figure 2.2). Having a common process view improves the relationships between collaborating organisations and helps project teams to plan operational activities with greater certainty and reliability. Together, this seeks to minimise the 'fuzzy edges between consultants and specialist engineering contractors' to which Latham refers.[16]

Collaborative strategies can be constructed between design chain members by understanding the flow of design information between them. Latham made the point that, if individual design processes can be identified, the responsibility for completing them can be allocated between collaborating organisations. In fact, Latham made specific reference to a draft guidance note as an exemplar for the construction industry. His plea for similar documents in other fields of design, however, has not yet been taken up.[17]

[16] Latham, M. (1994) *Constructing the Team: Joint review of procurement and contractual arrangements in the United Kingdom construction industry*, HMSO, London.

[17] Parsloe, C.J. (1997) *Allocation of Design Responsibilities for Building Engineering Services (Technical Note TN21/97)*, Building Services Research and Information Association, Bracknell, UK . A companion volume (TN 22/97) gives exemplar drawings and documents to inform the allocation of design responsibilities.

Second ICD Principle: Adopting Supply Chain Management Practices

After applying process management, organisations adopting an ICD approach are ideally positioned to assemble a design chain from their supply chain relationships. The object is to make the sum greater than the parts and this is done by organisations integrating their design competencies with one another.

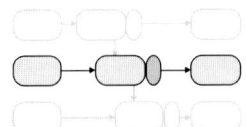

In ICD, design information is exchanged between project parties in a similar way to the transfer of goods within the supply chains of other industries.
Not surprisingly, ICD adopts many SCM practices. In applying these to construction, ICD makes an important distinction between supply chains and supply networks:

- **A supply network is a group of organisations with known competencies (technical and/or managerial) which have previously worked together and which exhibit a degree of mutual understanding that has arisen from past experience.**

- **A supply chain is a project-specific group of organisations consciously brought together to provide all the competencies (technical or managerial) required to complete a project. These are one-off arrangements largely, though not exclusively, drawn from a common supply network.**

ICD makes an important distinction between supply networks and supply chains.

SCM in ICD is the management of supply networks in the business domain and the selection of supply chains in the project domain.

The ICD distinction between supply chains and supply networks is a reflection of the project-based nature of the construction industry. In other industry sectors, such as fast-moving consumer goods, supply chains and supply networks are synonymous due to constant interaction associated with the delivery of a stream of goods. In construction, the flow of orders or contracts is intermittent; consequently, there is a clearer distinction between business and project relationships.

A design chain is a specialised form of supply chain and, by viewing a project as a design chain, the organisations within it can be managed according to whether their employees provide or receive design information. Thus, by viewing a construction project as a design chain and by identifying the design processes within it, ICD applies SCM principles to both the business domain and the project domain. In the business domain this means that ICD provides methods of selecting companies for collaborative working, while, in the project domain, it provides methods of identifying supply network members with the required technical and managerial competencies and the means to assemble them into project-specific supply (and design) chains.

It should also be noted that ICD is concerned primarily with the management of design information including site assembly information. The flow of goods or services between organisations during site assembly might also use SCM techniques; however, this does not fall within the remit of ICD.

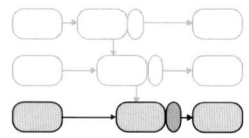

Third ICD Principle: Establishing Value Frameworks

The final stage in the evolution of an organisation's approach to ICD is concerned with establishing value frameworks. These build stronger business domain relationships with other organisations and allow project domain mechanisms to be used more effectively. These promote the delivery of value by using collaborative design. A variety of working methods combine to form a coherent approach to value delivery (Figure 2.3).

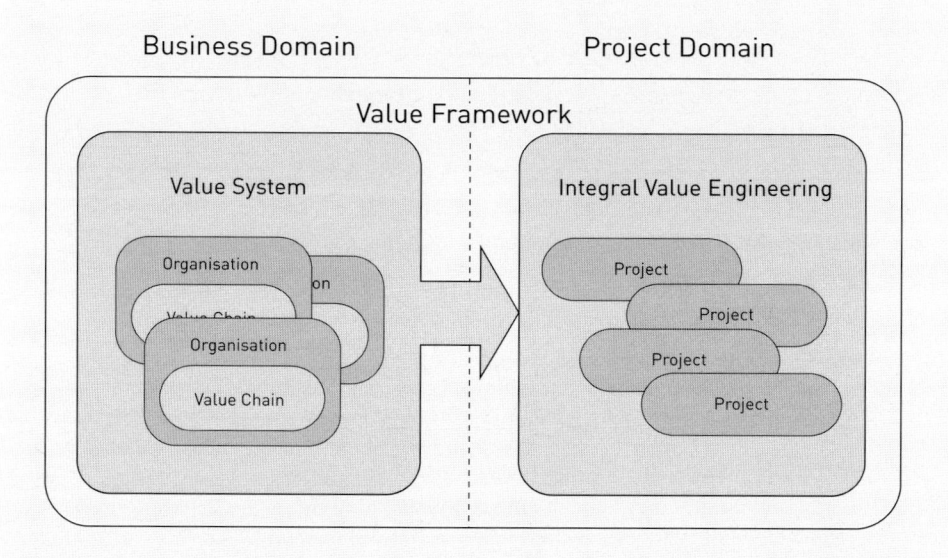

Figure 2.3: Relationships of value frameworks, value systems and value chains

The use of value chains to inform the sharing of processes between strategic partners is well established in many industries. Integrated partners may share their logistics, sales, IT infrastructure, or human resource management processes, for example. Porter,[18] Mintzberg et al.[19] and Johnson and Scholes[20] discuss this topic.

Building value frameworks begins with an organisation's value chains. These are used to review the organisation's internal processes to determine which contribute to project value. This is important because the better a company understands its own processes, the better it will be able to negotiate in sharing out the collaborative work. It is also important to identify any repeating processes in collaborative work, as these can be the basis of economies of scale.

The ICD value chain is used by organisations to structure the review of their internal processes. This helps them to build a value system out of an existing supply network by allowing them to determine which processes are suited to sharing with other organisations, and which are best retained. The value chain model ensures that (ideally) the core processes reflect each organisation's unique design expertise (Figure 2.4).

[18] Porter, M. E. (1985) *Competitive Advantage: Creating and sustaining superior performance*, The Free Press, New York.
[19] Mintzberg, H., et al. (1998) *The Strategy Process: Revised European Edition*, Pearson Education, Harlow, UK.
[20] Johnson, G., Scholes, K. (1999) *Exploring Corporate Strategy (5th Ed.)*, Pearson Education, Harlow, UK.

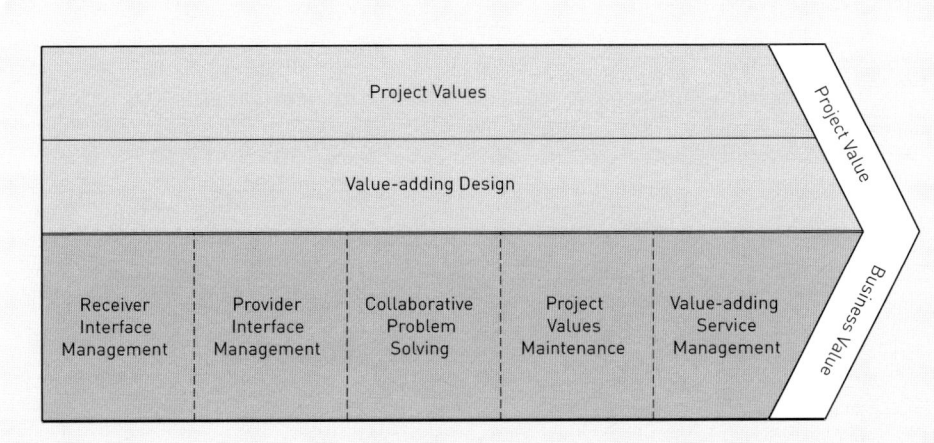

Figure 2.4:
The ICD value chain.
See Section 3.4 for
a detailed explanation

Value systems are built upon the relationships between organisations in supply networks. They are created by integrating the processes of several organisations and by rationalising common activities. This reduces overhead costs and improves the overall efficiency of the collaborating organisations. This integrated function is applied to projects through the project design chain. In the same manner as supply networks, value systems exist within an organisation's business domain.

Evidence of the effectiveness of value chains in aligning strategic business partners, and of value systems built upon supply chains, are commonplace in a number of industries. These rationalisations are concerned with the elimination of waste when physical items are transferred from one place to another. In the same way, ICD value systems are concerned with eliminating waste when design information is exchanged.

Decisions about the sharing of processes between organisations influence project domain activities but are made at a strategic level in the business domain. The sharing of non-value-adding (but still necessary) design tasks depends on the level of trust built into the supply network. This is because it contributes to the integration and, therefore, the interdependency of value system members. Hence, the ICD principle of establishing value frameworks helps organisations to integrate their common processes in the business domain to improve their efficiency in the project domain. This delivers business value to value system members and project value to both design chain members and to the client.

2.2 Elements of the ICD Approach

Figure 2.5 summarises the basic parts of ICD. It shows how an organisation's ICD approach to design management is structured into two domains (the business and the project) and into two roles (of the *provider* and the *receiver*).

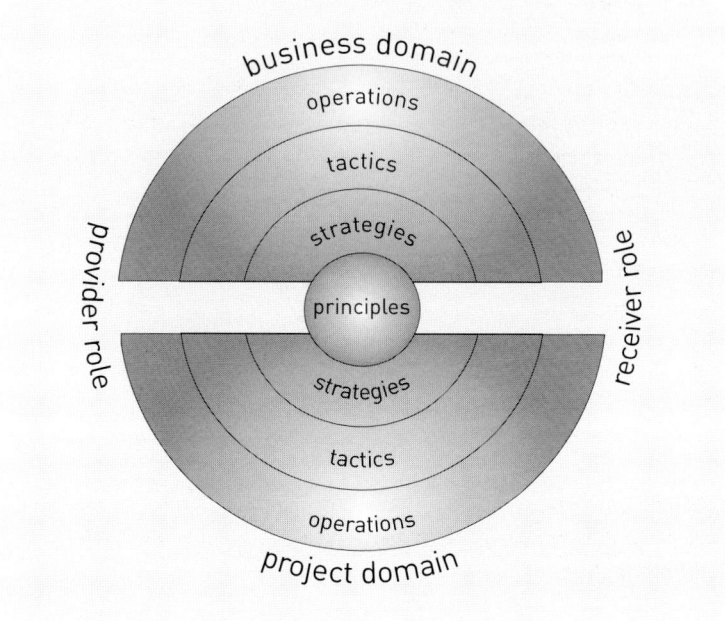

Figure 2.5: The structure of an ICD approach to design management

ICD Principles

At the core of ICD are the three constant principles of:

- **applying process management;**
- **adopting supply chain management practices; and**
- **establishing value frameworks.**

Once adopted by an organisation and integrated into its culture, they establish a common understanding and attitude among its members. This supports the use of all other elements of the ICD approach. The adoption of these principles is the necessary first step for collaborative working.

ICD Practices

ICD *practices* are categorised into strategies, tactics and operations. They have an order of precedence: from strategies, through tactics to operations. ICD strategies support an organisation's use of ICD tactics, which, in turn, support its application of ICD operations to its everyday work (Figure 2.6). This precedence is irrespective of whether the practice occurs in the business domain or in the project domain.

Part 5 of this handbook describes the ICD practices in detail.

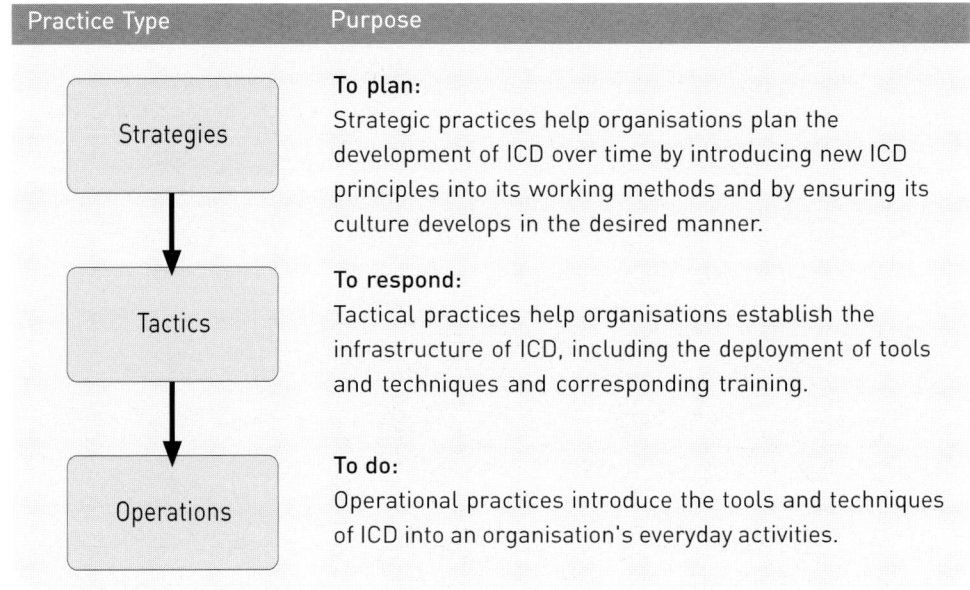

Practice Type	Purpose

Strategies

To plan:
Strategic practices help organisations plan the development of ICD over time by introducing new ICD principles into its working methods and by ensuring its culture develops in the desired manner.

Tactics

To respond:
Tactical practices help organisations establish the infrastructure of ICD, including the deployment of tools and techniques and corresponding training.

Operations

To do:
Operational practices introduce the tools and techniques of ICD into an organisation's everyday activities.

*Figure 2.6:
Different practice types build from one to the other in the deployment of an organisation's ICD approach*

ICD Roles

Within a design chain, an individual may be predominantly either a provider or a receiver of design solutions. These roles influence the manner in which ICD principles and practices are interpreted and applied.

ICD Domains

An organisation's approach to ICD influences its activities within both its business domain relationships (long-term) and its projects (short-term). To reflect this, ICD practices support both the business domain and the project domain (although individual practices are specific to one or the other).

The Difference between the Business Domain and the Project Domain

The ICD approach classifies individuals into two groups: those concerned with activity in the business domain and those concerned with activity in the project domain. Business domain operations are the ongoing activities of an organisation that give it structure. They span projects and establish the company in the market. Project domain operations, on the other hand, are temporary and occur when individuals and other resources come together to deliver individual projects. Examples of business domain activities include training, managing teams, client development, maintaining alliances, business development, and quality assurance; examples of project domain activities include project management, value management, design management and quality control.

The division between the business domain and the project domain is ambiguous. It is defined largely by the views that organisations and individuals have of themselves and of what they do. Some see the construction industry as being very task-focused, reflecting the emphasis placed on the highly technical and role-specific training that most individuals receive. However, this emphasis may cause managers and designers to concentrate on project details at the expense of their organisation's wider business.

Furthermore, the advent of the Private Finance Initiative, the Public Private Partnership and other long-term partnering arrangements are increasingly causing organisations to see major projects as separate business units.

Despite the subjective division between the business domain and the project domain, an ICD approach makes a distinction in order to highlight the importance of the planning and management of relationships that needs to take place in the business domain. It encourages the view of business as being concerned with the delivery of projects as an ongoing operation, rather than focussing on the isolated delivery of individual projects.

The following sections describe the business domain and the project domain in more detail and illustrate the importance of achieving an appropriate balance between them in an ICD approach.

The ICD Business Domain

Business domain activities are clearly vital to the survival of all organisations. The extent to which they are formally defined is dependent on the nature of the organisation. In general, larger organisations are more formalised and may have allocated responsibility for certain processes to dedicated staff, although the processes themselves are generic across all organisation types. Some organisations may consider some of their processes to be more important than others due to the nature of their work and the markets they serve.

A balance must be struck between the relationships formed in the business domain and in the project domain. Traditionally, relationships are formed on individual projects and, therefore, have been a function of the project domain. This has caused organisations to give insufficient consideration to managing and sustaining their long-term relationships in the business domain. This is not conducive to design chain operation.

An ICD approach structures the otherwise ad hoc management of those business domain activities concerned with developing and maintaining an organisation's long-term relationships. For example, a receiver, having developed a relationship with a provider in the project domain, may be influenced by the performance of that provider's organisation and promote it for inclusion in the supply chain of subsequent projects. Currently, methods of capturing, filtering and maintaining this knowledge are rarely developed much beyond simple databases or networks of personal relationships. Project domain relationships provide weak feedback for business domain relationships. An ICD approach addresses this by allowing business domain relationships to lead, driving project domain relationships on the basis of knowing the capability and the performance of business partners (Figure 2.7).

The ICD Project Domain

All organisations repeatedly apply a core set of processes to the projects they undertake. Irrespective of the nature of the work, many of these processes remain constant and are concerned with issues such as programme management, procurement, document control and resources. These processes have been extensively described and modelled by initiatives such as the *Plan of Work* published by the Royal Institute of British Architects[21] and the Process Protocol.[22]

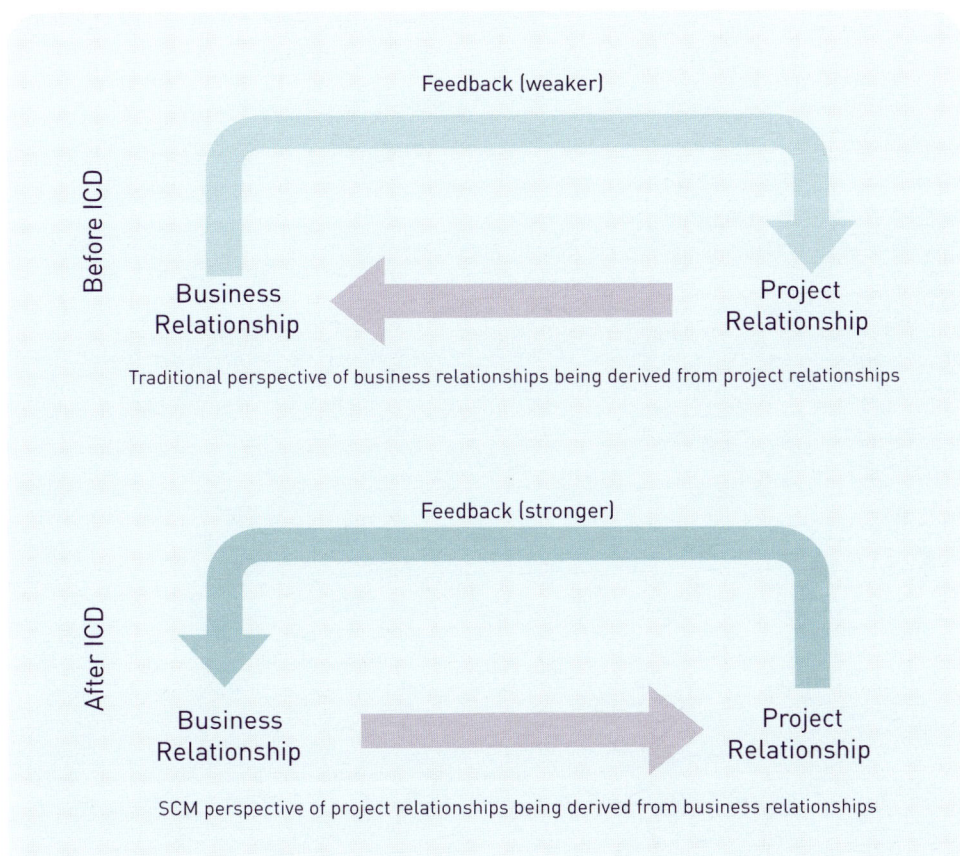

Figure 2.7:
The relationships of
projects and business
can be viewed in
different ways

The recently developed *Generic Design and Construction Process Protocol* has sought to establish a generic process map for projects that can be used as a framework for a common project process and language between collaborators (Figure 2.8). The Process Protocol provides a project management framework that helps organisations understand each other and work together better. This work on generic project processes has informed much recent work in supply chain integration by companies such as BAA, who have adopted framework partnership arrangements.

ICD is concerned with the control of complex design and design-management processes, and particularly with how they affect the design process. It addresses these issues with techniques such as mapping the flow of design information at the project level and by creating supportive business-to-business environments for value-adding collaborative work. Design chains are built by focusing on the business domain processes relevant to collaborative design. They help to provide a stable environment around projects and to inform the decision-making.

[21] Royal Institute of British Architects (RIBA) (2000) *The Architect's Plan of Work for the Procurement of Feasibility Studies, A Fully Designed Building Project, Employer's Requirements or Contractor's Proposals*, RIBA Publications, London.

[22] The Process Protocol is a process-based definition of documentation and procedures to help organisations involved in a construction project work together seamlessly. See www.processprotocol.com for more information.

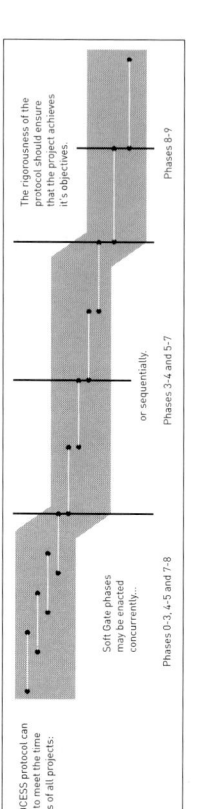

Figure 2.8:
The Process Protocol map

2.3 ICD Practices

Strategies

Part of the strategic planning that needs to take place to use ICD must include establishing working relationships that promote the use of collaborative design. Unlike general business strategies, ICD strategies are not concerned with diversifying an organisation's core competencies nor with developing new ones. Instead, they are concerned with managing the manner by which its current competencies are linked with the other organisations with which it works.

ICD strategies are divided into those concerned with an organisation's business domain activities and those concerned with its project domain activities (Table 2.1).

	Business domain strategies	Project domain strategies
Applying Process Management	**Planning Design Process Management (BS1)** To give ICD organisations the opportunity to capture, communicate and maintain a generic representation of their business and project processes. This will help them understand the tasks that must be performed and identify who is responsible.	**Planning Project Design Management (PS1)** To plan how design management practices are deployed within the project domain and to ensure design chain organisations operate in a manner that satisfies their business objectives. Organisations will be able to understand and improve their project and design processes.
Adopting Supply Chain Management Practices	**Planning Supply Chain Management Business Practice (BS2)** Describes a range of issues that influence how an organisation needs to plan the way it implements SCM.	**Planning Supply Chain Management Project Practice (PS2)** To develop strategies to ensure that all projects are undertaken according to the SCM strategy of the organisation defined in its business domain. This includes assembling supply network members to form a project design chain.
Establishing Value Frameworks	**Planning the Implementation of Integral Value Engineering across the Business (BS3)** To plan the development of an organisation's Integral Value Engineering (IVE) practice within its established practices and long-term objectives.	**Planning Integral Value Engineering Project Practice (PS3)** To ensure that all IVE practice planning, which takes place in the project domain, is consistent with the strategic planning that has taken place in the business domain. This includes managing the provision of IVE resources within projects for shared use within a design chain.

The ICD research project has developed a range of tools and techiques that can support integrated collaborative design. These are put forward in this handbook as the ICD Practices. They should not be seen as the only tools that are available, just those from this project. Organisations adopting an ICD approach should also bring to bear other methods which they find useful and complementary.

Table 2.1: ICD strategic practices

Tactics

As with ICD strategies, tactics are divided into those that support long-term business development in the business domain and those better suited to short-term project environments in the project domain (Table 2.2).

	Business Domain Tactics	Project Domain Tactics
Applying Process Management	**Applying Process Management in the Business (BT1)** To provide a generic framework of processes encompassing all types of activities performed within a project, throughout its duration from the clients' conception of a business need through to asset realisation.	**Applying Design Management Practices (PT1)** To develop approaches for the application of additional design management operations (including ADePT, process modelling and DePlan) which are provided by ICD and intended to work within a collaborative environment.
Adopting Supply Chain Management Practices	**Aligning Supply Networks (BT2)** Aligns the business and the project processes of supply network members to allow organisations to work together as part of a design chain. **Applying Supply Chain Management in the Business (BT3)** To help companies establish Supply Chain Management working practices within their ICD approach, which structure their use of supply networks to inform the assembly of project supply chains and, therefore, design chains.	**Applying Supply Chain Management to a Project (PT2)** To plan how design chains are to be assembled from an existing supply network and, then, how the chain will be used to deliver a project. **Selecting Supply Chain Members at the Project Level (PT3)** To make use of the technical design competencies that exist within the supply network and, more specifically a design chain, which has one or more specialised skills they can contribute to projects.
Establishing Value Frameworks	**Applying Integral Value Engineering in the Business (BT4)** To manage the use of IVE to ensure that it corresponds with business objectives and make available the resources required for projects. **Conducting a Value Survey (BT5)** A value survey is used by a receiver to gather a snapshot of the methods used by providers (typically within a value system) to address value within their work. **Performing an ADePT Review (BT6)** Seeks to identify generic groups of design activities to focus the use of IVE to address the underlying design complexity.	**Implementing Integral Value Engineering on a Project (PT4)** This practice forms a bridge between tactical responses to business strategies concerned with the development of an organisation's, or a value system's, use of IVE resources on projects.

Table 2.2:
The ICD tactical
practices

Operations

A variety of operations introduce ICD principles to everyday business domain and project domain activities. Some represent new activities for which new skills must be learnt, while others comprise design management methods already practised elsewhere but which can be applied within the context of the ICD.

ICD business domain operations are concerned with advancing an organisation's strategy. Part of this process involves reviewing the feedback from projects and, where necessary, making appropriate strategic adjustments to incorporate new knowledge derived from projects. The use of operations in the business domain and the project domain is the same: to provide a response to the need to achieve strategic objectives (Table 2.3).

	Business Domain Operations	Project Domain Operations
Applying Process Management	**Modelling Business Processes (BO1)** To model complex business design processes in a simplified format. This helps organisations align their external interfaces with internal functions by communicating their desired design scope and information requirements to others.	**Applying ADePT to Design Management (PO1)** To use ADePT to identify the optimal sequence of design activities to satisfy the development of a design solution, and develop detailed design programmes that take into account the iteration inherent within the design processes. **Applying DePlan to Design Management (PO2)** To ensure that providers and receivers manage the design process in a transparent environment, where supply chain organisations are aligned, and information transfer is streamlined. To improve the completion of design to programme by focusing on the removal of constraints and weekly scheduling of achievable tasks. **Modelling Project Design Processes (PO3)** To model project design operations in a simplified way and help align a project design chain. This involves analysing the business design project models available in the design chain and developing an appropriate model for the specific project.

Table 2.3:
ICD Operational Practices
(continued overleaf)

Table 2.3: continued

	Business Domain Operations	Project Domain Operations
Adopting Supply Chain Management Practices	**Auditing the Supply Network (B02)** To establish the SCM competencies present within a supply network and to identify, for each member, its working practices that give rise to these competencies. **Auditing the Supply Network for Technical Competence (B03)** To establish the technical capabilities present in organisations who may be a part of a supply network. The audit benchmarks the organisation against minimum required levels of performance for network membership and information for the selection of design chain members against technical project criteria.	**Assembling Supply Chains (P04)** To form selected organisations of known competency into a supply and design chain to structure their involvement in projects.
Establishing Value Frameworks	**Gathering Value-adding Feedback from Projects (B04)** To gather evidence of value-adding tool application on projects and determine their effectiveness in adding value. Feedback is collected on the way value-adding tools have been used for inclusion in a value-adding toolbox and to provide an audit trail.	**Applying Value-adding Tools to Design Problems (P05)** To support the application of value-adding tools to design problems so that problem-solving frameworks can be created to solve technical design problems while also addressing project values.

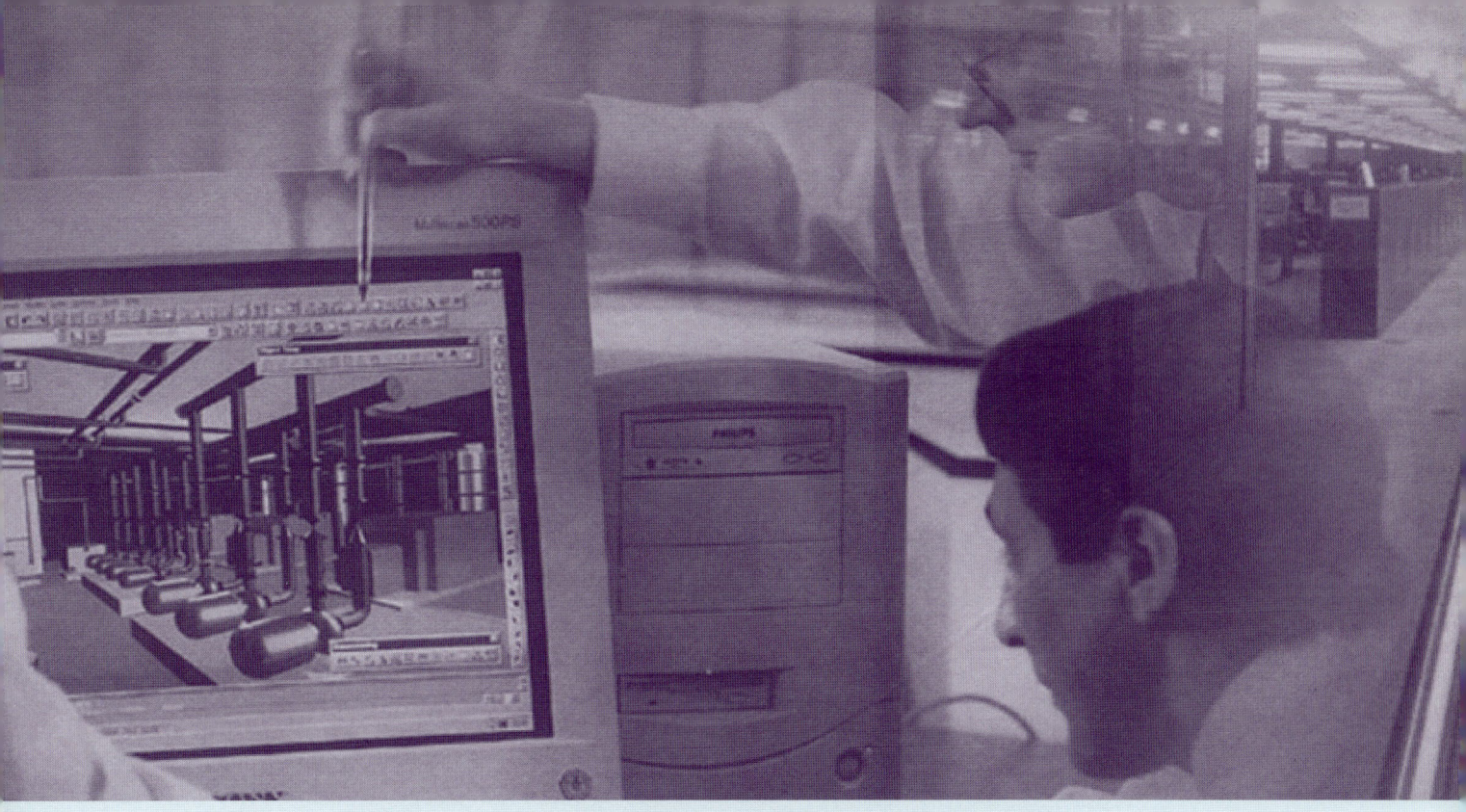

Part 3: Applying ICD Principles

3.1 Advancing an ICD Approach

The ICD terminology is explained in the Glossary at the back of this handbook.

The purpose of this part of the handbook is to help you to progressively adopt and apply the three ICD *principles*. We show the relationship between their application in the *business domain* and the *project domain* and outline the main steps required. You will then be in a position to start using the supporting tools and techniques that comprise the ICD *practices*, described in Part 5.

By adopting each of the three ICD principles in turn (Figure 3.1), an organisation can:

- use design process modelling to understand design information flows and allocate design tasks between organisations appropriately;

- build on its process models by establishing supply networks to group together organisations of known technical competency, so that design responsibilities can be allocated among them; and

- integrate the processes of organisations within the supply network to build a value system.

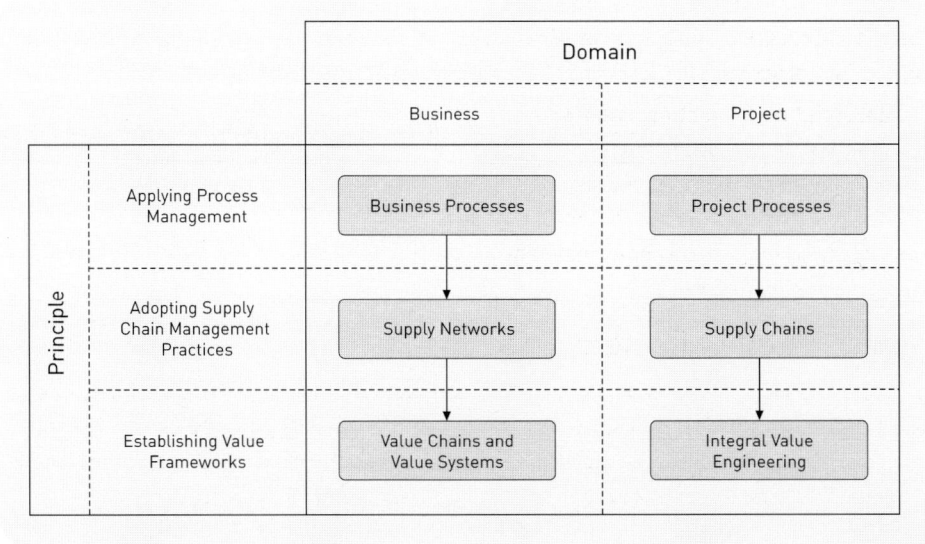

Figure 3.1: Advancing an ICD approach within an organisation's business domain and project domain

This approach means that practice improvements progress from one stage of business domain ICD development to the next, rather than from one project to the next. For example, in the project domain, supply chains are not dependent upon project processes in the same way that supply networks are dependent upon business processes. This part of the handbook helps organisations understand these principles and the manner by which they build from one to the other so that they can apply them effectively.

3.2 How to Apply Process Management

Viewing projects in terms of their constituent processes helps organisations increase their shared understanding of business domain practices. This allows the scope of work to be well defined, roles and responsibilities agreed by all, and information transfer to be managed to meet project, as opposed to business, needs. Traditionally, the flow of this information has proved difficult to identify and to manage. *Providers* typically push too much information without first determining exactly what is required. *Receivers* then tend to be swamped with information they may not need. Time is wasted in deciphering and sorting through information supplied to determine whether it contains the necessary elements and, if not, requesting what is missing. This wastes money and effort, which could be avoided if providers and receivers understood each other's information needs.

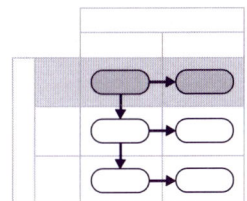

To overcome this, the management of design information flow must look beyond being concerned with ensuring that providers push all possible information to the receivers to seeing that receivers have the ability to pull just the information they need.

Design chains are built and operated to simplify organisational relationships. These relationships are built in the business domain and are then applied in the project domain. This design activity can only be aligned by establishing a detailed understanding of who is to do what. The following six steps are necessary to be able to manage this process:

1 **understanding project processes;**
2 **modelling project processes;**
3 **aligning organisation interfaces;**
4 **enhancing design information co-ordination;**
5 **establishing project transparency; and**
6 **fostering project learning.**

Understanding Project Processes

A project comprises a series of processes to achieve a defined objective. Design is one of these processes and a step towards the project solution (Figure 3.2). However, design information flows are complex and iterative. This makes them difficult to manage and it can be hard to spot when a final design solution has been completed.

Figure 3.2:
A simple process view –
information is required to
develop a solution

Individual organisations are usually familiar with their internal processes but have less understanding of how these relate to the processes of others. Traditionally, organisations have relied upon contracts to define how they interact with each other on projects but they are not the most effective way of ensuring collaboration within a design chain. Organisations need to remain flexible. This is best achieved by allocating responsibility for project processes on the basis of competencies.

Modelling Project Processes

To model processes, organisations need to define the project processes for which they are responsible. There are many different ways to do this. The method used must let organisations capture a true representation of their processes and the links these have with other chain members; graphical representation is a method many favour.

The first step in producing a model is to look at the structure of an organisation's work processes, starting with the highest level and then descending to identify all process systems and sub-systems (Figure 3.3). These structures can be reviewed by all organisations in a design chain before starting a project. By doing this, the design tasks performed by each organisation can be aligned within long-term business domain relationships to ensure that there are no gaps or overlaps in the way the processes have been allocated.

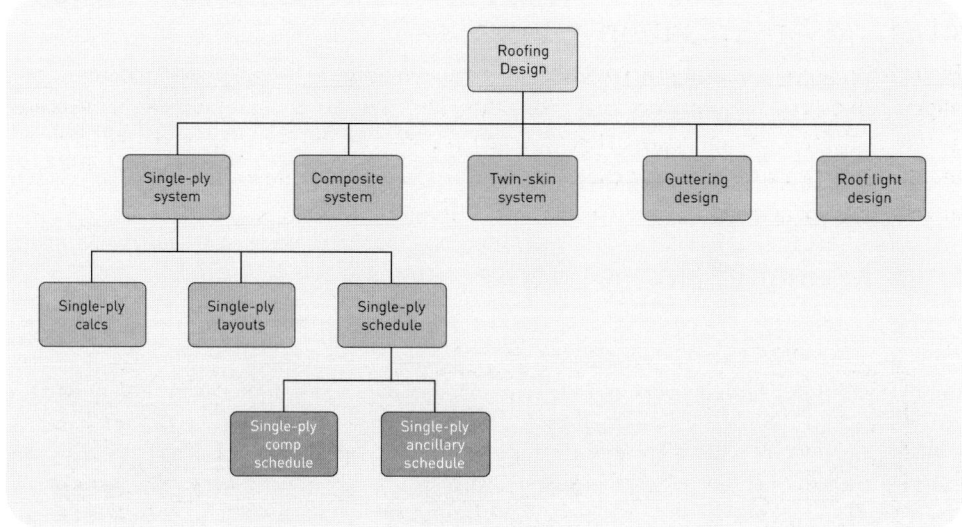

Figure 3.3:
Example work breakdown
structure, providing the
basis for process
modelling and task
allocation between
organisations

To provide sufficient detail to allocate design processes effectively, each organisation's work breakdown structure should be accompanied by a detailed design process model illustrating the flow of information between individual design tasks.

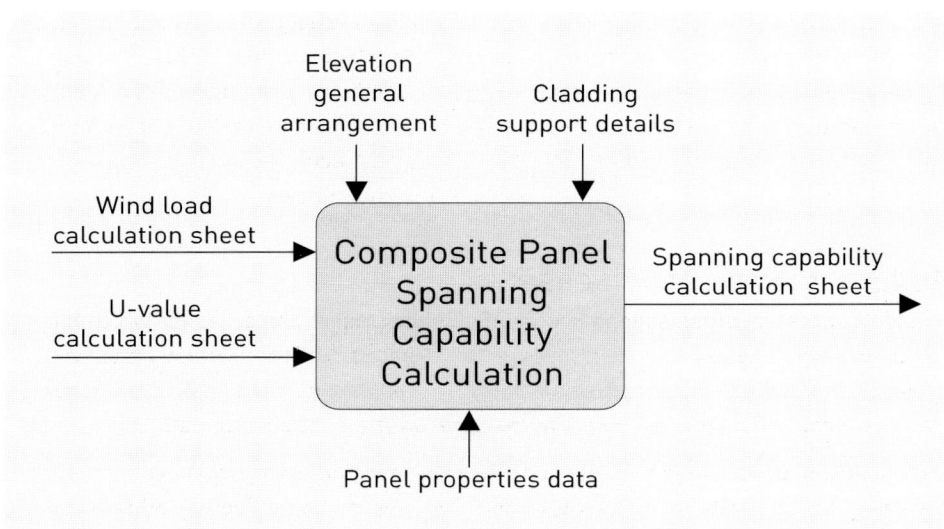

Figure 3.4:
Modelling a detailed level
design task by
representing information
flow using the IDEF0v
notation[23]

By understanding the flow of information within projects, design activity can be divided into its constituent processes, according to the flow of information between them. Graphical notations, such as data flow diagrams and IDEF0[24] are suitable for this purpose (Figure 3.4).

Aligning Organisation Interfaces

By defining project design processes, the interfaces between organisations can be aligned on a project-by-project basis. This simplifies design management, as it becomes easier to allocate design tasks between organisations and makes for a seamless transfer of information. It also becomes easier to verify that no gaps or duplications in the information flow between tasks have occurred (Figure 3.5).

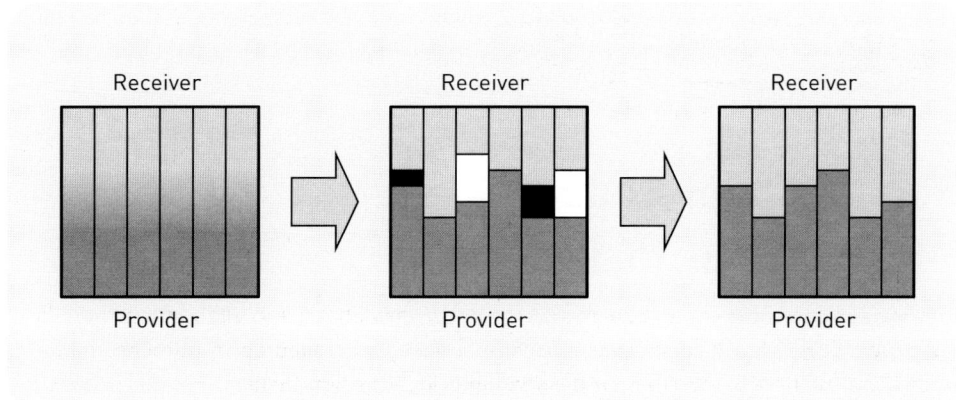

Figure 3.5:
Aligning organisations
through viewing projects
as processes

---

[23] Austin, S., Baldwin, A., Li, B. and Waskett, P. (1999) 'Analytical Design Planning Technique: A model of the detailed building design process', *Design Studies*, Vol. 20, Issue 3, April, pp 279-296.

[24] IDEF (Integrated Computer Aided Manufacture Definition) is a suite of modelling techniques developed by the US Air Force.

Furthermore, project procurement can be improved by timing the release of design information to co-ordinate with the physical production of components, delivery to site, and final assembly.

Enhancing Design Information Co-ordination

Design co-ordination has historically been problematic. Typically, this is due to a lack of understanding of the information requirements of project parties. This tends to cause one organisation to push design information onto another, irrespective of the requirements of the receiving organisation. Design chains, on the other hand, encourage the pull of design information from one process to the next. This information pull may occur between design processes within a single organisation or between design chain members. To pull information from providers, receivers must communicate their needs, explaining what is required and why it is important.

Establishing Project Transparency

Viewing projects in terms of their constituent processes helps ICD practitioners to operate in a transparent environment with well-defined and mutually agreed roles and responsibilities and where information transfer is managed to meet project needs as opposed to business needs.

Transparency may be difficult to establish if processes are poorly understood. Without transparency, organisations cannot communicate effectively, reducing the likelihood of integrated project solutions being achieved and making it difficult to determine project progress. This, in turn, makes for unreliable planning, can contribute to wasted effort and resources and can cause projects to overrun.

Fostering Project Learning

As organisations repeatedly work with each other, they learn from their experiences of managing design collaboratively. To improve their combined effectiveness, they can feed project lessons learned back to the business domain from where they can be re-applied to subsequent projects.

3.3 How to Adopt Supply Chain Management Practices

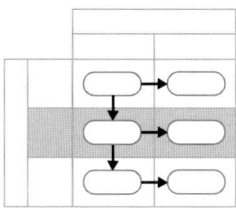

Supply chain management (SCM) differs from many other management concepts in the degree to which it is the product of evolving industry practice. SCM has emerged through the application of a variety of management tools and techniques, which together have encouraged managers to view the delivery of products or services as the result of co-ordinated action by groups of organisations. While people can see the benefits of SCM for one industry (for example, the car industry[25]) not all can see that these are obviously transferable to construction.

---

[25] Ohno, T. (1988) *Toyota Production System: Beyond large-scale production*, Productivity Press, Cambridge, Mass.

A better way to view SCM is to see it less as a particular technique but more as a framework that brings together a variety of independent practices. Within this framework SCM gives these practices direction and equips their users with the necessary shared values to apply them in a coherent fashion. This is the approach ICD takes to SCM, which can be adopted by following six elements:

1 establishing business networks;
2 building business relationships;
3 adopting a holistic approach to product or service delivery;
4 changing attitudes;
5 linking business SCM with project SCM; and
6 creating design chains from supply chains.

Establishing Business Networks

The term 'supply chain' describes a series of linear links. The chain refers only to those organisations involved with a particular product, project or service that is sold to the client or user who is at the end of the chain. The ICD approach distinguishes between the 'supply chain' - for individual projects with clear beginnings and ends - and the 'supply network', which is a group of organisations that repeatedly work together on different projects. With ICD, it is from 'supply networks' that 'supply chains' are assembled (Figure 3.6). The supply network is used to form the long-term, strategic relationships between organisations necessary as a precursor to practising SCM.

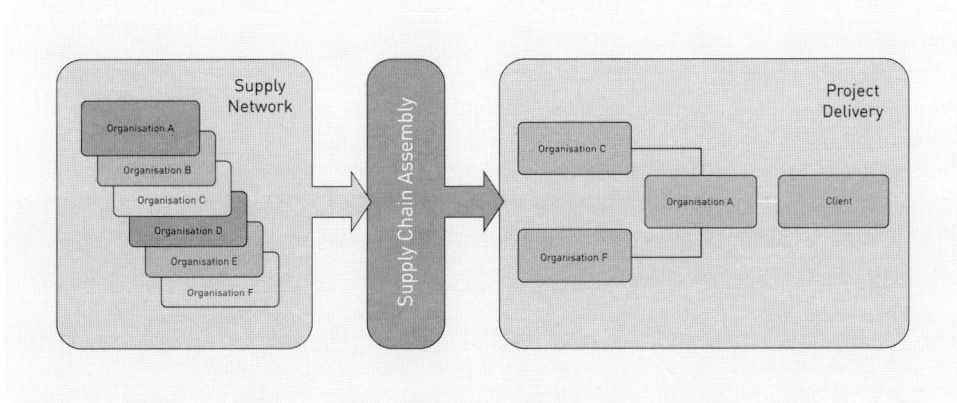

Figure 3.6:
Relationship of supply
network and supply chain

Building Business Relationships

Supply networks exist in the business domain and are relationships that bind organisations together. These relationships can be both vertical (between different levels of providers and receivers) and horizontal (between organisations on the same tier). They arise because few organisations, by themselves, can satisfy all the demands of their customers. New technologies, the need to optimise product delivery and shortening programmes, are examples of the forces causing organisations to work together in order to survive.

Individual companies can survive by collaborating in complex arrangements but where each retains ownership of some core competencies. In this arrangement, each organisation can become increasingly specialised in its core competencies to the benefit of the whole and each can collaborate in the management of an external network.

Adopting a Holistic Approach to Product or Service Delivery

The concern with the organisational relationships within the business domain means that SCM is concerned with looking beyond an organisation's short-term interest in the supply of a product or service. An organisation applying SCM practices will shift its focus and become concerned with how its performance affects the ability of the chain to deliver. Responsibility for the overall performance of the supply chain lies with all members of the chain, although the ability to influence that performance will differ between organisations or design disciplines.

Changing Attitudes

The importance that relationships play in an organisation's continued success means that it is no longer appropriate to let them be formed on an ad-hoc basis. Relationships need to be managed and to be encouraged to move forward from construction's long-held adversarial culture. Many new concepts, tools, techniques and strategies are being developed to help organisations work together. ICD brings several of these together into a coherent approach.

Attitudes influence the types of relationship organisations form and business practices (such as payment arrangements, contracts, dispute resolution procedures) all have an impact on how organisations relate to each other. Table 3.1 compares traits of traditional project domain relationships with those required for SCM and which are embodied in the ICD approach.

Collaborative Project Domain Relationship	Traditional Project Domain Relationship
Proactive	Reactive
Co-operative	Competitive
Trust	Distrust
Two-way information	One-way information
Mutual obligation	Contractual obligation
Honour-bound to repay	Take advantage
Long-term focus	Short-term focus
Interdependence	Independence
Co-destiny	Survival

Table 3.1:
Characteristics of
collaborative and
traditional project
relationships

Linking Business SCM with Project SCM

Building business relationships between organisations requires a managed approach. In most cases, the better managed a relationship is the more effective it will be. This means companies benefit from fewer but deeper relationships.

Some project-level processes can help with the management of relationships in the business domain. They provide the means to:

- develop generic project processes;
- deepen the understanding of the technical competencies of other organisations; and
- anticipate where the interfaces between organisations on projects will occur.

Creating Design Chains from Supply Chains

Just as we have seen that supply chains are project-specific groupings of organisations, we can view design chains as a specialised form of supply chain. Within an ICD approach, a design chain is a sub-set of the project supply chain. It involves only those companies that contribute to the design. This is because many of the specialist practices used to deliver design solutions (such as process modelling and integral value engineering) are not applicable to the wider project supply chain. Within design chains, individuals perform the roles of providers and receivers of design solutions.

3.4 How to Establish Value Frameworks

ICD value frameworks, which integrate processes from different organisations, provide business domain benefits. There are three components to such frameworks:

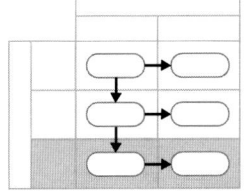

1 A value chain is a device used to examine and classify an organisation's internal value-related processes. It exists in the business domain.

2 A value system is built upon a supply network to integrate business partners. It also exists in the business domain and delivers business value.

3 Integral Value Engineering (IVE) is the implementation of design to deliver value-adding solutions. It exists in the project domain and delivers project value.

Value systems are built from the relationships that exist in supply networks between the members. The trust established between these organisations provides the basis for the integrating of business and project processes between organisations. There are six elements to building value frameworks:

1 understanding business value and project value;
2 value system prerequisites;
3 examining business processes using a value chain;
4 building a value system from a supply network;
5 introducing value-system benefits to projects; and
6 practising IVE.

Understanding Business Value and Project Value

Value systems are driven by the need to improve the value which organisations derive from their long-term relationships in the business domain. The 'business value' they create differs from 'project value' in that it benefits the organisation while project value benefits the project team and the client. However, the use of value systems to generate business value can indirectly improve projects because they create relationships and frameworks that are more conducive to collaborative work.

Value System Prerequisites

The ability of an organisation to join a value system is dependent on its having previously adopted other ICD principles, namely:

- **understanding, mapping and allocating responsibility for common design processes between value system members - the Applying Process Management ICD principle; and**

- **to stabilise business relationships and move beyond competitive working, the organisation must have also adopted the ICD principle of Adopting Supply Chain Management Processes.**

Appropriate ICD practices associated with both of these also need to be integrated into working methods.

Examining Business Processes using a Value Chain

The use of value chains is a well-established business management technique. They were originally developed to classify and examine the internal business processes of an organisation. From this, organisations can identify which of their internal processes are suitable for sharing with strategic partners (Figure 3.7). They are especially useful in collaborative ventures where a number of partners produce different elements of a complex product and its selling price is affected by how efficiently it can be assembled.

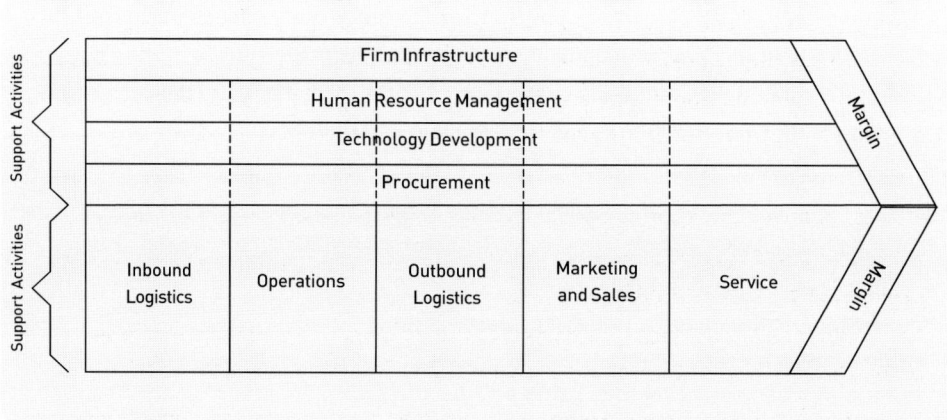

Figure 3.7: Original generic value chain structure, as proposed by Porter

The Porter[26] value chain has been adapted for use within ICD (Figure 3.8). It is used to classify an organisation's processes according to their ability to deliver internal business value through the organisation's involvement in project design chains.

---

[26] Porter, M. E. (1985) *Competitive Advantage: Creating and sustaining superior performance*, The Free Press, New York.

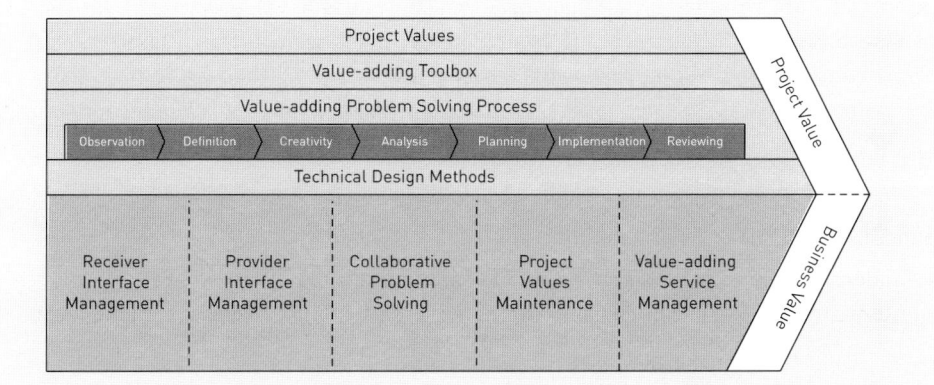

Figure 3.8:
The ICD value chain -
adapted from Porter

In ICD, the value chain model represents generically the key value-related internal processes of organisations. The processes are developed and managed in the business domain and they are applied to all projects that the organisation undertakes. By optimising these processes the organisation can not only generate business value for itself but can improve the performance of the design chain of which it is a part and, hence, derive project value.

Building a Value System from a Supply Network

The establishment of a value system starts with individual supply network members using the ICD value chain to examine their internal processes to see where efficiencies can be created and which processes are suited to being shared. The sharing of common design processes reduces the complexity of design information flow and increases the combined effectiveness of the organisations involved.

Shared processes tend to be of a facilitating nature. The entire value system, rather than individual system members, is optimised through a process of negotiation to improve the efficiency of value-adding design. As discussed above, the mechanisms used to share processes emerge over time and are built on pre-existing trust. Within a value system, individual members may adopt sub-optimal solutions so that the process of allocation and harmonisation throughout the system can be mutually beneficial to all its members.

The ability of an organisation to engage in a value system is dependent on its:

- **prior use of SCM to create and sustain the operation of business supply networks and project supply chains; and its**

- **harmonisation of business strategies and the establishment of sufficient trust between partnered organisations to begin to integrate duplicated processes.**

By repeatedly working with each other, network members develop an understanding of each other's processes and, eventually, network members can begin to propose processes for sharing with other members - thus, starting the value system. Depending on the stage of its development, a supply network may be a 'simple' supply network (with co-ordinated, pre-qualified but segregated members) or it may have a complete, or partially complete, value system built upon it. Given that a value system is built upon a supply network, the relationships of these two elements of ICD practice must be clearly understood (see Figure 3.1).

The role of supply networks in an ICD approach is described in Section 3.3.

Supply networks and value systems represent stages in the development of the relationships between partnered organisations. Irrespective of how mature the ICD implementation is, a supply chain is always used to draw together the technical competencies from either a supply network or a value system[27] for application to the project domain.

If design activity alone is considered, the design chain emerges as a sub-set of the project supply chain. Instead of building directly upon each other, as in the business domain, the elements of value-related ICD practice used in the project domain complement, rather than replace, those developed at the previous stage of the practice. Within a value system, one of the shared processes is the maintenance of mechanisms that can be used to deliver project values by supporting the practice of integral-value engineering in projects. Within an ICD approach, two mechanisms lie at the core of project value delivery:

1 a portfolio of value-adding tools, which are assembled to form processes by which technical design solutions can be structured and collaboration promoted; and

2 a value-adding toolbox, which provides a mechanism to disseminate techniques and lessons learnt from their previous projects (value-adding tools).

Introducing Value System Benefits to Projects

It can be seen from Figure 3.9 that the first tier provider (who may be a lead project contractor, lead consultant or a project management organisation) channels the value-adding contributions of all project members to the client. This is done by creating the interface through which project value expectations are identified, understood and responded to by the design chain. Value management, which is used initially to do this, is commonly applied to the early, conceptual stages of project.

The role of the organisation forming the primary interface between the project and the procuring client (a design management contractor, lead consultant, or a project manager, for example) is analogous to the distribution and retail channels in manufacturing. This organisation must understand the customer's needs and see that they are met. To ensure project success, the unique requirements of each end-user must be understood and an appropriate project solution derived. The technical and process competencies required to deliver the solution must also be understood. If a value system has begun to be built upon a supply network, then the formation of a project supply chain will also align the value chains of individual value system members. This creates an environment in which the project value-related practices (in Part 5) can be deployed to form project design chains.

The ICD practices (in Part 5) associated with the application of SCM practices provide methods of selecting organisations for supply network membership and for the formation of supply chains.

---

[27] Assuming organisations have been audited for SCM maturity.

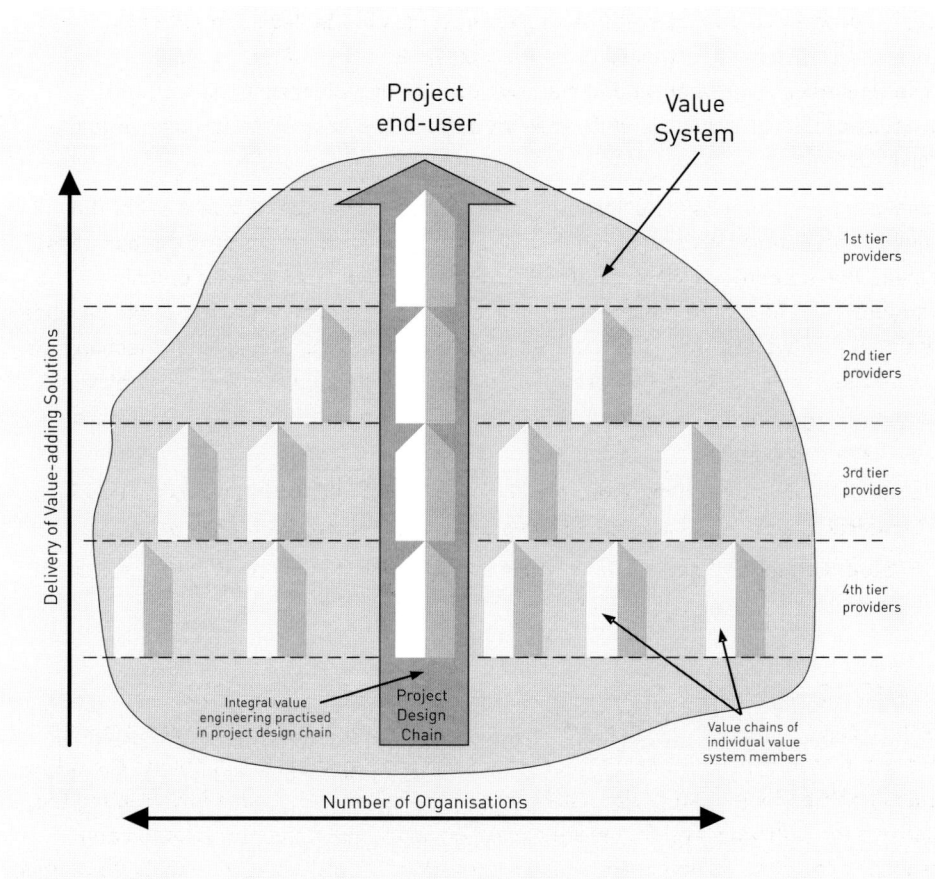

Figure 3.9:
Relationship of business
value system to project
design chain

Practising Integral Value Engineering

Integral Value Engineering (IVE), used for delivering project values, provides the framework for the use of a variety of mechanisms to devise design solutions that will satisfy client values and, therefore, project stakeholders.

IVE is used to address the relationship between project design solutions and the relevant project values they must satisfy (defined by the traditional use of value management at the outset of the project). It is this problem-solving approach, using value-adding tools to create processes, which promotes collaboration. It relies on the trust and willingness of value system members to share processes in the business domain. Although IVE can be practised without a supply network, it will be less effective at promoting collaborative working because the levels of trust will be lower than with a harmonised value system.

3.5 Assessing your Progress with ICD Principles

Having reviewed the structure and content of the three ICD principles, it is recommended that you assess how they are reflected in your organisation's current practice. Most business will have embarked on establishing elements of ICD practice in response to the consumer and market pressures outlined in Section 1.2. Several techniques can be used to carry out this assessment of where you are against these recommendations, from a maturity analysis as indicated in Tables 3.2, 3.3 and 3.4 to a simple assessment of current practice against each of the sub-sections of the ICD principles.

This assessment is intended primarily to be a navigational aid for readers of this handbook, although it can be used as the basis for a more rigorous assessment performed individually or within a group to assess the status of ICD within a team or an organisation. For example, detailed assessment techniques, based on the European Foundation for Quality Management 'Excellence Model', could be applied on an annual basis to help prioritise ICD development. Each assessment table has six rows, representing the six elements of the ICD principles. The table columns represent the stages of maturity of an organisation, working from left to right, following the structure of the CBPP Construction Performance Driver.[28]

To determine the status of each principle within an organisation, its corresponding table should be reviewed and the appropriate level identified for each element. Each table should then be reviewed in its entirety and an overall judgement made regarding the status of the principle within your organisation, based on the level of its elements. The tables to help structure your assessment are on the following pages.

---

[28] BQC Performance Management Ltd and Construction Best Practice Programme (2001) *The Construction Performance Driver: A health check for your business*, BQC Performance Management Ltd in partnership with the Construction Best Practice Programme.

Maturity Assessment of Applying Process Management

	Level 1 Don't know	Level 2 Haven't thought about it	Level 3 Thinking of doing something about it	Level 4 Doing it as normal business	Level 5 Full deployment and improvements	Level 6 Inherent practice throughout operations
Understanding Project Processes	No understanding of project processes ☐	Projects use processes defined by contract ☐	Recognise that existing processes need development ☐	Seeking alignment of project processes ☐	Coherent project process well established between organisations ☐	Consistent project process developed and aligned with competencies ☐
Modelling Project Processes	No knowledge of process modelling ☐	Individual businesses define own processes ☐	Consistent work breakdown structures recognised ☐	Generic processes established and used ☐	Shared work breakdown structures are understood and utilised ☐	Processes modelled and tasks aligned ☐
Aligning Organisational Interfaces	No attempts to define interfaces ☐	Inconsistent allocation of tasks with significant interface problems ☐	Design overlaps and gaps recognised in critical areas ☐	Task allocation between all organisations based on known interfaces and competencies ☐	Seamless transfer of information without gaps or overlaps ☐	Fully aligned interfaces with a just-in-time release of information ☐
Enhancing Design Information Co-ordination	Design information exchange is not effectively managed or monitored ☐	Design information is 'pushed' to all parties indiscriminately, regardless of need based on contracts ☐	Recognised overload of information flow and need to realign practices ☐	Design information shared in common format with clear understanding of needs ☐	Information needs of each organisation understood with predominantly 'pull' transfers of essential information only ☐	Fully co-ordinated needs expressed including 'what' and 'why' it is important ☐
Establishing Project Transparency	Silo mentalities predominate ☐	Obscured roles and poor communication cause significant delays to projects ☐	Communication problems recognised as cause of poor performance on projects ☐	Project roles are well defined and communicated ☐	Mutual agreement of roles is established as basis for co-ordination ☐	Fully effective communication based upon clarity of roles and responsibilities ☐
Fostering Project Learning	No acknowledgement of the impact of projects on delivery ☐	Problems recur on successive projects causing poor performance and delays ☐	Mechanisms exist for capturing feedback from individual projects ☐	Feedback from projects is consistently captured and shared ☐	Feedback is managed collaboratively across projects ☐	Project-based learning is fed back to business relationships to improve future project performance ☐

Table 3.2: Assessment of Applying Process Management Principle Maturity

Maturity Assessment of Adopting Supply Chain Management Practices

	Level 1 Don't know	Level 2 Haven't thought about it	Level 3 Thinking of doing something about it	Level 4 Doing it as normal business	Level 5 Full deployment and improvements	Level 6 Inherent practice throughout operations
Establish Business Networks	No awareness of any business networks ☐	Informal networks of individuals and organisations arising form project relationships ☐	Existence of informal networks is registered ☐	Informal networks are being developed into formal relationships ☐	Assessment or benchmarking of membership of formalised networks ☐	Stable networks facilitating co-development and joint - improvement ☐
Building Business Relationships	No recognition of the need for or extent of business relationships ☐	All relationships mediated through current projects ☐	Recognise existence of long-term informal relationships between individuals or organisations ☐	Attempts made to manage long-term relationships between individual organisations ☐	Development of formal structures to manage strategic relationships ☐	Structured approach to strategic relationships involving regular assessment and benchmarking performance ☐
Adopting a Holistic Approach to Delivery	Organisations operate solely to comply with contractual obligations ☐	Organisations operate independently to optimise their own performance ☐	Recognition that project success requires teamwork across organisations ☐	Interdependence resulting from longer-term relationships used to optimise delivery of combined services ☐	Formal arrangements between organisations created to facilitate integrated project approach ☐	Supply chain seen as unitary structure with organisational interests subordinate to overall service delivery. ☐
Changing attitudes	Defensive attitudes pervade; organisations seek to keep their distance from others ☐	Isolated pockets of co-operation dependent on personal relationships ☐	Development of implicit mutual obligations between businesses ☐	Changes to formal arrangements; project and business practices alongside developing trust ☐	Development of a hierarchy of organisational status amongst suppliers and customers ☐	Developed obligational relationships with long term, co-operative focus ☐
Linked Business SCM with Project SCM	No recognised link between business and project activities ☐	Project SCM activities predominate with weak links to business relationships ☐	Attempting to develop generic approaches to SCM project tasks ☐	Collaboration with individual organisations to improve project SCM practices between projects ☐	Use of network as a resource to develop generic project practices and project arrangements ☐	Integrated network acts as a virtual organisation ☐
Creating Design Chains from Supply Chains	No awareness of the existence of or need for design chains ☐	Contractual management of design input provided by different organisations ☐	Inconsistent approach to managing design on a project by project basis ☐	Recognition of the role of design in project supply chain ☐	Practices developed to integrate design capability of networked organisations ☐	Delivery of integrated design chains providing integrated design solutions ☐

Table 3.3: Assessment of Adopting Supply Chain Management Principle Maturity

Maturity Assessment of Establishing Value Frameworks

	Level 1	Level 2	Level 3	Level 4	Level 5	Level 6
	Don't know	Haven't thought about it	Thinking of doing something about it	Doing it as normal business	Full deployment and improvements	Inherent practice throughout operations
Understanding business value and project value	No recognition of difference between business and project value ☐	The importance of managing business and project activities according to role in creating value not appreciated ☐	Recognise that there might be a difference ☐	The role of individual processes in contributing to business or project value is understood ☐	Process management reflects their role in creating business or project value ☐	Processes are optimised according to their role in creating business or project value ☐
Value system prerequisites	Not aware of the status of process modelling and SCM within organisation ☐	Organisation operates traditionally, without process modelling or SCM ☐	Aware that process modelling and SCM are required to build value frameworks ☐	Process modelling and SCM are being established ☐	Process modelling and SCM fully established ☐	Process modelling and SCM are an inherent part of how 'things are done' ☐
Examining business processes using a value chain	Not aware of existence of value chain ☐	Business processes are optimised according to information flow alone, without regard to their potential value-adding role ☐	The ICD value chain is recognised as suitable device to examine value-adding role of processes ☐	The ICD value chain is used to classify processes according to role in provision of business or project value ☐	Detailed understanding of role of processes in creating business and project value ☐	Management of processes reflects their importance in creating business and project value ☐
Building a value system from a supply network	Not aware of what a value system is or the business benefits it could bring ☐	Supply network members work in isolation from each other and do not relate their activities ☐	Recognition that the creation of a value system would bring mutual benefit to supply network members ☐	Supply network members have identified common processes and are beginning to share them ☐	Value system members are mutually interdependent due to the sharing of common processes ☐	Value system members offer services as a single business ☐
Introducing value system benefits to projects	Value system exists in the business domain, but methods of applying integrated working to projects not yet established ☐	No evidence of a value system in the business or project domain ☐	Opportunity to improve effectiveness of project design chain using value system relationships identified ☐	Value chains used routinely within organisations to identify processes suited to sharing in value system ☐	Design chain members align their internal processes using value chains prior to projects commencing ☐	Co-ordinated interface to client for all design chain members ☐
Practising integral value engineering	Design chain exists, but members do not co-ordinate their project activities ☐	No evidence of IVE practice or resources ☐	Infrastructure for IVE in place, but yet to be fully used ☐	Project design solutions developed using collaborative working between value system members ☐	IVE resources shared between all design chain and value system members ☐	Projects deliver fully-integrated solutions that make the best use of all design chain members' expertise ☐

Table 3.4: Assessment of Establishing Value Frameworks Principle Maturity

3.6 Putting ICD Principles into Practice

So far, we have reviewed the structure and discussed the three ICD principles that must be in place for collaborative working:

1 **applying process management;**
2 **adopting supply chain management; and**
3 **establishing value frameworks.**

Once adopted and established within the corporate values of an organisation, these principles can be put into operation by using a variety of ICD practices. These are documented in Part 5 of this handbook. How to determine which ICD practices are relevant to you and your organisation is discussed in the next section.

Now that you have determined the extent to which each of the ICD principles are embedded in the way your organisation does things, you can progress to the selection of suitable ICD practices for use in your daily work. The following Part 4 will guide you through this.

Part 4: Applying ICD Practices

4.1 Introduction

The purpose of this part of the handbook is to help you to identify appropriate ICD *practices*. This depends on the stage of adoption of the three ICD principles within your organisation (Section 4.2). You also need to know:

- your primary responsibilities in your organisation, whether they are associated with *business domain* or *project domain* activities (Section 4.3);

- type of practice you wish to apply to your organisation's activities - this can be at the strategic, tactical, or operational level (Section 4.4); and

- your role in your organisation, as predominately either a *provider* or a *receiver* of design information (Section 4.5).

You need to know the extent to which the ICD principles have been adopted by your organisation because if these principles are only partially established, then the use of ICD practices should be restricted to an appropriate sub-set.

> ℹ️ *The ICD terminology is explained in the Glossary at the back of this handbook.*

In this section you will be presented with background information describing relevant characteristics of the domain, deployment level, stage of principle adoption and roles. You will then be guided on how to determine which practices are appropriate (Figure 4.1). These are divided into six categories:

- **Business Strategies**
- **Business Tactics**
- **Business Operations**

- **Project Strategies**
- **Project Tactics**
- **Project Operations**

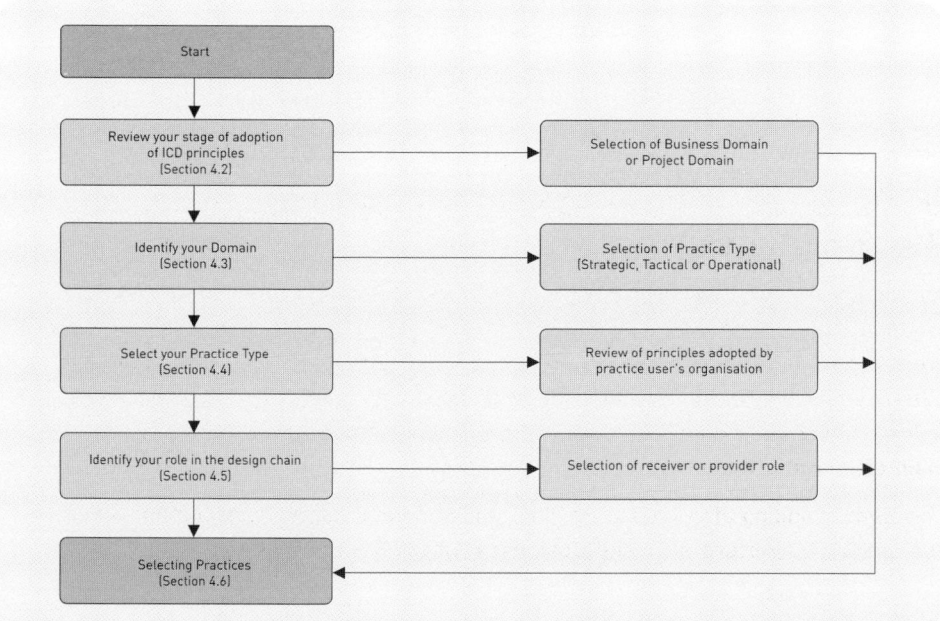

Figure 4.1:
Practice selection
process

4.2 Review your Stage of Adoption of ICD Principles

This stage of the selection process reviews the extent to which our organisation has adopted each of the three ICD principles (described in Part 3). This review is necessary because several ICD practices assume a prior understanding of the ICD principles.

You should now review the profile of your organisation generated by performing the assessment recommended in Section 3.5. While it is recognised that some organisations will already exhibit elements of all three principles, it is important that the strategy to fully adopt the principles takes place in the correct sequence. In other words, you should focus first on achieving full maturity in applying process management, then adopting SCM practices and finally establishing value frameworks (because, recalling Section 2.1, they build upon one another). For example, if your organisation has made progress with regard to SCM but has not addressed process management, you should not concentrate on the former ahead of the latter, although some degree of overlap is inevitable.

Summary

Record here which of the ICD principles is the current focus of your organisation:

ICD Principle	Tick
1. Applying Process Management	☐
2. Adopting Supply Chain Management Principles	☐
3. Establishing Value Frameworks	☐

4.3 Identify your Domain

Understanding Processes in the Business Domain and the Project Domain

An understanding of business domain and project domain processes underpins an organisation's ICD approach. As an organisation begins to understand the way it works in process terms, it begins to identify areas where it has a comparative advantage over other companies. Having identified its advantages, an organisation then needs to apply them to the projects on which it works (and to constantly improve those processes to maintain any advantage).

This understanding of processes is also fundamental to other management concepts such as Quality Assurance[29] and Business Process Re-engineering.[30]

The principle of adopting a process view of construction is discussed in Sections 2.1 and 3.2.

---

[29]Thorpe, B., Sumner, P., Duncan, J. (1996) *Quality Assurance in Construction* (2nd ed), Gower, Aldershot, UK.
[30]Champy, James (1995) *Reengineering Management*, HarperCollins Publishers, London.

ICD uses this understanding as the means to increase an organisation's awareness of the activities it performs, be they design, production, assembly, or the selection and classification of organisations for collaborative working.

Through mapping business domain processes and project domain processes, organisations can better understand the relationship between the two. Organisations are commonly involved in several projects at any one time, each of which may be at a different stage of delivery. As each project progresses through the business (Figure 4.2), it interacts with the ongoing business domain processes in the same way that a raw material may physically move through a factory being operated on at each stage of manufacture. Just as organisations use factories to provide stability for the physical environment of manufacturing, organisations can use their business domain processes to provide economic, social and organisational stability around projects. This frees project staff to concentrate on the management of the project processes and the effective delivery of the project. In this way the business domain acts as a filter lying between the project and the wider business concerns.[31]

> *Organisations can use their business domain processes to provide economic, social and organisational stability around projects.*

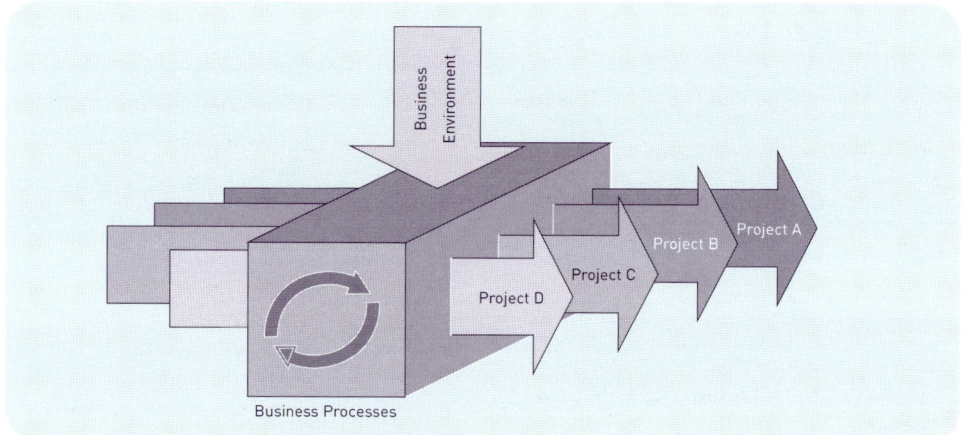

Figure 4.2:
The movement of projects through the permanent organisation's business processes

An example of this stabilising effect is the agreement of annual fixed prices with materials suppliers in the business domain. This can have the effect of introducing supply price stability into the project domain. Without these agreements, the specialist contractor may experience variations in the product price between one project and another. Through the business process of agreeing annual prices, project stability is provided by insulating projects from the wider business environment. Similarly, in an ICD approach, design processes are modelled and agreed in the business domain. This provides the basis for allocating responsibility for performing design tasks between strategic partners, providing stability to the projects they undertake together and minimising the possibilities of both duplication and omission.

---

[31] Some lower tier organisations in the supply chain may already have a strong focus on the business domain, with relatively weak interfaces with the project domain.

Although the ICD approach differentiates the business domain from the project domain, these different types of activity are not necessarily performed by different groups of people. Managers are usually involved in both, albeit with distinct roles. As you read this, you should be aware of whether your own priorities lie predominately in the business domain or the project domain.

Your Involvement in Business Domain Activities

To determine the extent to which you are concerned with business activities, you should review the extent to which you create working relationships within single projects and the business through which projects are delivered. Examples of business domain activities include:

- marketing;
- client development;
- nurturing and maintaining strategic business relationships;
- monitoring supply network performance;
- managing the value system and monitoring performance;
- managing teams;
- provision of business resources;
- business process management and improvement;
- knowledge management;
- quality assurance and TQM;
- developing standards; and
- training and education.

Your Involvement in Project Domain Activities

You are concerned with project domain activities if you are involved with the delivery of either a single or several concurrent projects. The act of design is a project domain activity, but developing a generic design guide is a business domain activity since it is not project-specific. Other examples of project domain activities include:

- project management;
- value management;
- design management;
- understanding specific client business needs;
- developing project responses to needs;
- assembling project supply chains;
- practising integral value engineering;
- planning and managing design tasks;
- allocating responsibilities for design tasks within the design chain;
- completing design tasks;
- constructing and delivering projects;
- managing project teams; and
- quality control.

You should now be in a position to assess where you feel most comfortable within your organisation - whether you see yourself as being concerned primarily with the project domain or with the wider business domain.

4.4 Select your Practice Type

To ensure that you select the right type of ICD practices for your requirements, you need to decide how you will be applying them to your organisation's activities. As discussed in Section 2.2, ICD practices can be deployed to serve strategic, tactical or operational needs. The type of practice relevant to your needs is determined largely by your function within your organisation.

The following boxes are to record the type of ICD practice you think is appropriate to your requirements (tick one box only).

	Tick one box
Strategic Practices - To plan: Strategic practices help organisations plan the development of ICD over time by introducing new ICD principles into their working methods and by ensuring their culture develops in a conducive manner.	☐
Tactical Practices - To respond: Tactical practices help organisations establish the necessary ICD infrastructure, including the deployment of tools and techniques and training their employees.	☐
Operational Practices - To do: Operational practices introduce ICD tools and techniques into an organisation's everyday activities.	☐

4.5 Identify your Role in the Design Chain

Construction organisations use a variety of labels to describe themselves: main contractor, consultant, specialist contractor and specialist supplier are common examples. These labels often reflect traditional roles, they may not take account of how each organisation's role has evolved over time. These labels can reflect stereotypical images that rarely match reality. Anecdotal evidence suggests that the term 'supplier' in SCM is objected to by design consultants and by specialist contractors. As elsewhere, labels in construction confer status. It may be subjective, but the term sub-contractor is seen by many to be so low down the hierarchy that many who perform this function now call themselves specialist contractors. For ICD, labels and relabelling causes problems when trying to apply a generic approach across a wide range of different types of organisation and roles in construction.

To overcome this problem ICD uses the terms provider and receiver to describe the roles of individuals in the exchange of design information throughout the design chain. Figure 4.3 illustrates the relationship between a provider and a receiver across any tier of the design chain.

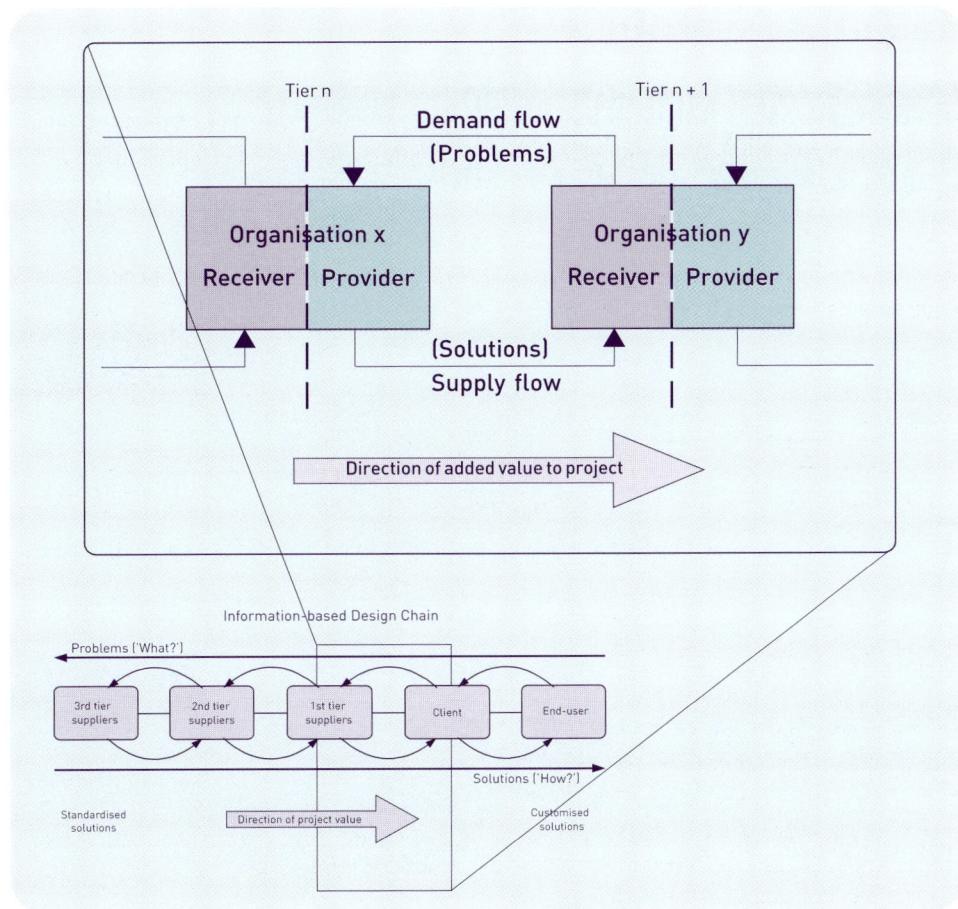

Figure 4.3:
Positions of receiver and
provider across a tier of
the design chain (see
Figure 1.2 also)

Design, as discussed previously, is inherently iterative and, in this respect, is quite different from a product-based supply chain. This makes it difficult to classify the position or tier where any type of organisation lies in a supply chain because individual employees may be acting as a provider or receiver. What you as an individual need to know is your role in exchanging design solutions with others as this determines whether you are primarily a receiver or a provider.

It can be seen that your role depends on whether you are looking up or down the chain - in terms of the organisations with which you are dealing - and, therefore, whether you are predominantly concerned with setting problems (a receiver) or generating solutions (a provider).

> By this point you should be able to assess whether your role is primarily a receiver or a provider of design solutions.

4.6 Selecting Practices

Descriptions of these practices are provided in Part 5

This section will guide your selection of relevant ICD practices. For each set of decisions a group of practices is recommended. Once the appropriate group has been identified, all the practices within it should be reviewed. From each group, those capable of supporting your current requirements should be selected for use.

Recall your assessment in Section 4.2 and note its outcome below:

	Tick one box
My organisation is focusing on the principle of applying process management	☐
My organisation is focusing on the principle of adopting supply chain management practices	☐
My organisation is focusing on the principle of establishing value frameworks	☐

You can now identify the appropriate practice selection table to your needs.

Recall your assessment in Section 4.3 and note its outcome below:

	Tick one box
I am concerned primarily with business domain activities	☐
I am concerned primarily with project domain activities	☐

Recall your assessment in Section 4.4 and note its outcome below:

	Tick one box
I require strategic practices	☐
I require tactical practices	☐
I require operational practices	☐

Recall your assessment in Section 4.5 and note its outcome below:

	Tick one box
I will use the practice(s) in a provider role	☐
I will use the practice(s) in a receiver role	☐

You should now look at the practices available to you in the appropriate table overleaf. You can review the purpose of each practice by reference to the summary in Tables 2.1 to 2.3 or to the individual practice description in Part 5.

	Strategic practices	Tactical practices	Operational practices
Applying process management	**Business practices:**		
	Planning Design Process Management (BS1)	Applying Process Management in the Business (BT1)	Modelling Business Processes (BO1)
	Project practices:		
	Planning Project Design Management (PS1)	Applying Design Management Practices (PT1)	Applying ADePT to Design Management (PO1) Applying DePlan to Design Management (PO2) Modelling Project Design Processes (PO3)

Adopting Supply Chain Management practices

Establishing value frameworks

Table 4.1:
Practices for applying
process management

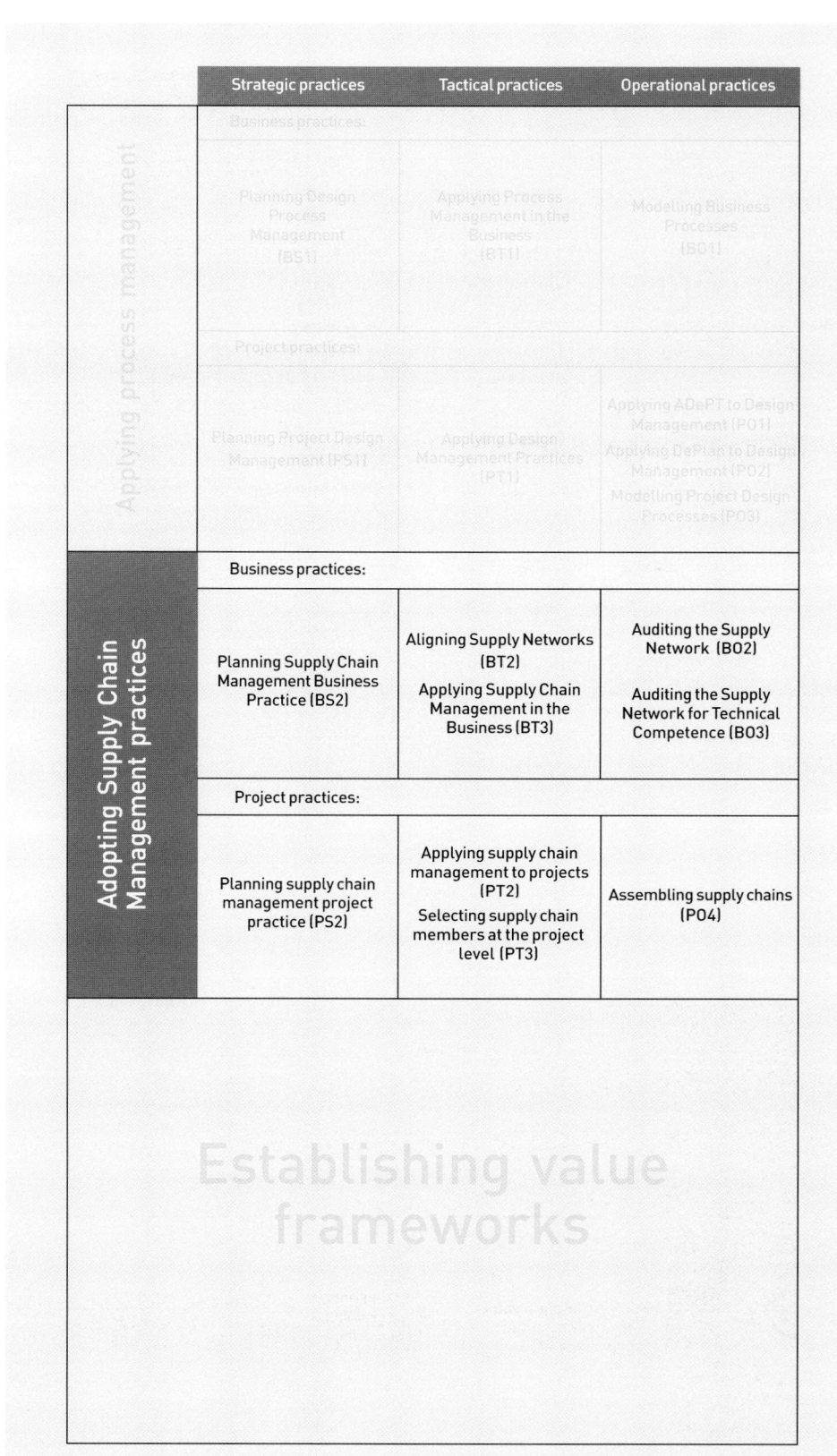

	Strategic practices	Tactical practices	Operational practices
Applying process management Business practices:	Planning Design Process Management (BS1)	Applying Process Management in the Business (BT1)	Modelling Business Processes (BO1)
Project practices:	Planning Project Design Management (PS1)	Applying Design Management Practices (PT1)	Applying ADePT to Design Management (PO1) Applying DePlan to Design Management (PO2) Modelling Project Design Processes (PO3)
Adopting Supply Chain Management practices Business practices:	Planning Supply Chain Management Business Practice (BS2)	Aligning Supply Networks (BT2) Applying Supply Chain Management in the Business (BT3)	Auditing the Supply Network (BO2) Auditing the Supply Network for Technical Competence (BO3)
Project practices:	Planning supply chain management project practice (PS2)	Applying supply chain management to projects (PT2) Selecting supply chain members at the project level (PT3)	Assembling supply chains (PO4)
Establishing value frameworks			

Table 4.2:
Practices for
adopting SCM
practices

	Strategic practices	Tactical practices	Operational practices
Applying process management — Business practices:	Planning Design Process Management (BS1)	Applying Process Management in the Business (BT1)	Modelling Business Processes (BO1)
Applying process management — Project practices:	Planning Project Design Management (PS1)	Applying Design Management Practices (PT1)	Applying ADePT to Design Management (PO1) Applying DePlan to Design Management (PO2) Modelling Project Design Processes (PO3)
Adopting Supply Chain Management practices — Business practices:	Planning Supply Chain Management Business Practice (BS2)	Aligning Supply Networks (BT2) Applying Supply Chain Management in the Business (BT3)	Auditing the Supply Network (BO2) Auditing the Supply Network for Technical Competence (BO3)
Adopting Supply Chain Management practices — Project practices:	Planning supply chain management project practice (PS2)	Applying supply chain management to projects (PT2) Selecting supply chain members at the project level (PT3)	Assembling supply chains (PO4)
Establishing value frameworks — Business practices:	Planning the Implementation of Integral Value Engineering across the Business (BS3)	Applying Integral Value Engineering in the Business (BT4) Conducting a Value Survey (BT5) Performing an ADePT Review (BT6)	Gathering value-adding tool feedback from projects (BO4)
Establishing value frameworks — Project practices:	Planning integral value engineering project practice (PS3)	Implementing Integral Value Engineering on a Project (PT4)	Applying Value-adding Tools to Design Problems (PO5)

Table 4.3:
Practices for establishing value frameworks

The Essentials

What is a design chain?

A design chain is that part of a project supply chain function that is focused on design. A design chain is created when organisations, which repeatedly collaborate in project supply chains, co-ordinate their roles to develop project design information for their mutual benefit. To achieve this, design chain members make a series of provisions in both their long-term business domain relationships and their short-term project domain relationships. In the same way that supply chains have revolutionised the production of goods and services, design chains, which exist as separate processes carried out by different organisations, can be managed as a chain across organisational boundaries.

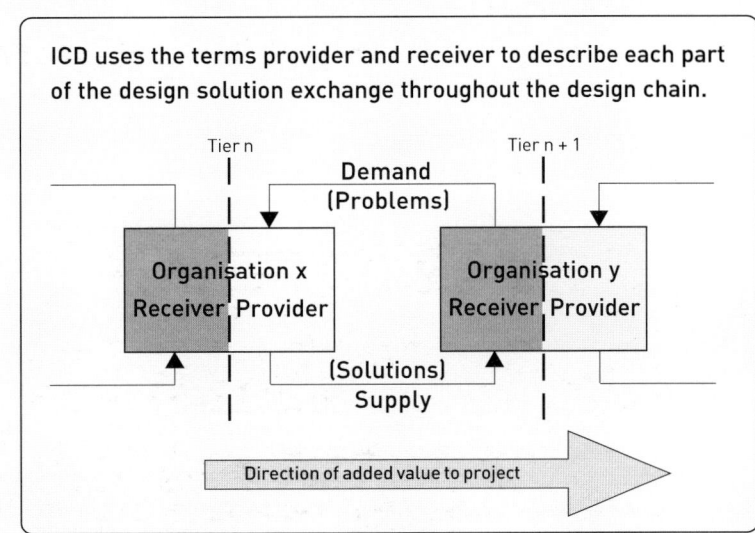

ICD uses the terms provider and receiver to describe each part of the design solution exchange throughout the design chain.

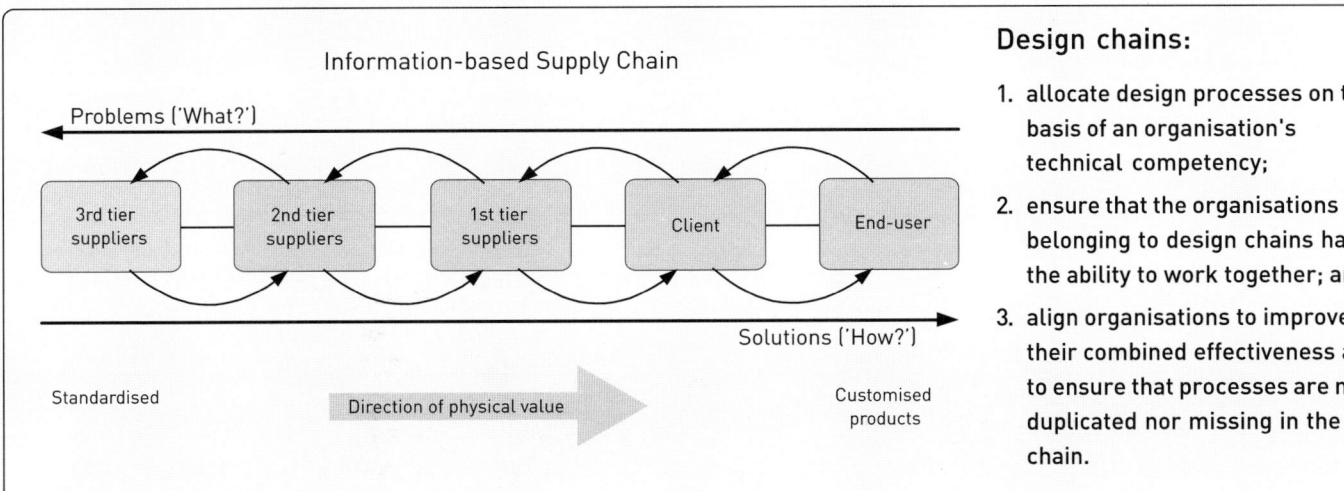

Design chains:

1. allocate design processes on the basis of an organisation's technical competency;

2. ensure that the organisations belonging to design chains have the ability to work together; and

3. align organisations to improve their combined effectiveness and to ensure that processes are not duplicated nor missing in the chain.

Design chains are concerned with the flow of design information between the organisations collaborating on a project. If differs from a supply chain in not having information flowing in one direction and material in the other. In a design chain, information flows both ways. However, the ICD design chain distinguishes between two types of information: problems (which frame requirements) and solutions (which provide a solution to a previously defined problem). For example, a performance specification may be compiled to frame the design problem for a lift installation. The specification defines the problem and the resulting lift design from the manufacturer is a solution to that problem. By understanding the relationship of design problems to design solutions, individuals can define themselves as either a provider or a receiver (or both) of design information for a particular level (tier) in the design chain.

As with the supply chain, where standard products are converted into bespoke facilities, solutions in the design chain become increasingly specialised and complex. Standard solutions (such as standard design details) are combined to provide more comprehensive and bespoke design solutions.

The Essentials

What is Integrated Collaborative Design?

ICD is an approach that establishes design as the common thread linking organisations together. At its core are a set of principles, which are applied to business domain and project domain activities and are supported by strategic, tactical and operational practices. It provides a basis for managing relationships between strategic partners, for building design chains and distinguishing between provider and receiver roles within the design chain.

ICD Principles

At the core of ICD are three principles that underpin an organisation's collaboration with others when devising design solutions:

1. **Adopting process management.**
2. **Applying supply chain management practices.**
3. **Establishing value frameworks.**

Once adopted by an organisation and integrated into its culture, they establish a common understanding and attitude among its members. This supports the use of all other elements of their ICD approach. The adoption of these principles is the necessary first step for collaborative working.

ICD Practices

ICD practices are categorised into strategies, tactics and operations. They have an order of precedence: from strategies, through tactics to operations. ICD strategies support an organisation's use of ICD tactics, which, in turn, support its application of ICD operations to its everyday work. This precedence is irrespective of whether the practice occurs in the business domain or in the project domain.

ICD Domains

An organisation's approach to ICD influences its activities within both its long-term business domain relationships and its short-term projects. To reflect this, ICD practices support both the business domain and the project domain. Business domain operations are the ongoing activities of an organisation that give it structure. They span across projects and establish the company in the market. Project domain operations on the other hand are temporary and occur when individuals and other resources come together to deliver individual projects.

Practice Type	Purpose
Strategies	**To plan:** Strategic practices help organisations plan the development of ICD over time by introducing new ICD principles into its working methods and by ensuring its culture develops in the desired manner.
Tactics	**To respond:** Tactical practices help organisations establish the infrastructure of ICD, including the deployment of tools and techniques and corresponding training.
Operations	**To do:** Operational practices introduce the tools and techniques of ICD into an organisation's everyday activities.

The Essentials

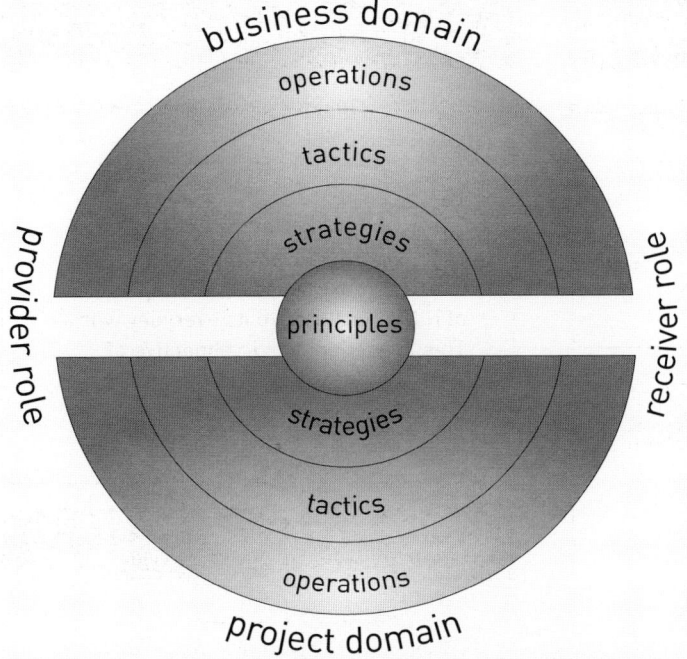

ICD Roles

Within a design chain, an individual may be predominantly either a provider or receiver of design solutions. These roles influence the manner in which ICD principles and practices are interpreted and applied.

Application

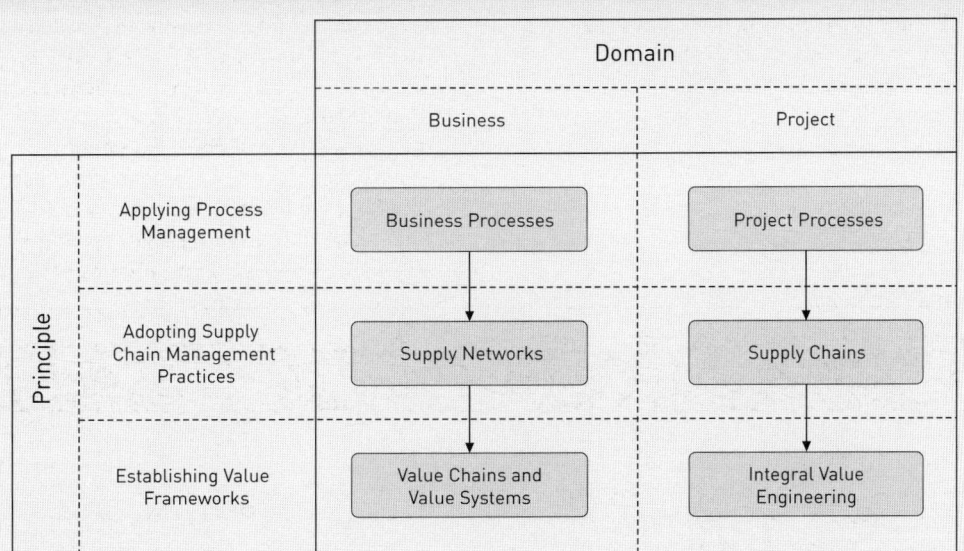

Application

By adopting each of the three ICD principles in turn, an organisation can:

1. use design process modelling to understand design information flows and allocate design tasks between organisations appropriately;

2. build on its process models by establishing supply networks to group together organisations of known technical competency, so that design responsibilities can be allocated among them; and

3. integrate the processes of organisations within the supply network to build a value system.

The Essentials

Benefits
The benefits of an ICD approach include:

- creating business level frameworks for organisations to collaborate with mutual benefit on projects;
- improving the understanding of information flow;
- helping organisations align their business and project competencies and cultures;
- promoting value-adding design solutions on projects; and
- providing a selection of tools to integrate processes across organisations at the levels of the business and project.

Relevance
ICD addresses issues of managing projects and creating business environments within which projects are delivered. It is relevant to all individuals and organisations that want to collaborate:

- in design;
- when procuring design;
- to manage the design process; and
- when applying design solutions.

Applying Process Management
By emphasising the generic nature of many design production tasks, organisations can manage their internal operations and interfaces with others to work collaboratively. Through modelling design processes and identifying overlaps and gaps in information flows between organisations, design activities can be allocated effectively and according to competency.

Adopting Supply Chain Management Practices
A supply chain is formed when organisations within a supply network of known competency collaborate on a project. By forming supply networks, organisations can improve business relationships and create an overarching business environment conducive to delivering integrated design solutions. Over time, networks promote trust between organisations and act as a forum for continuous improvement.

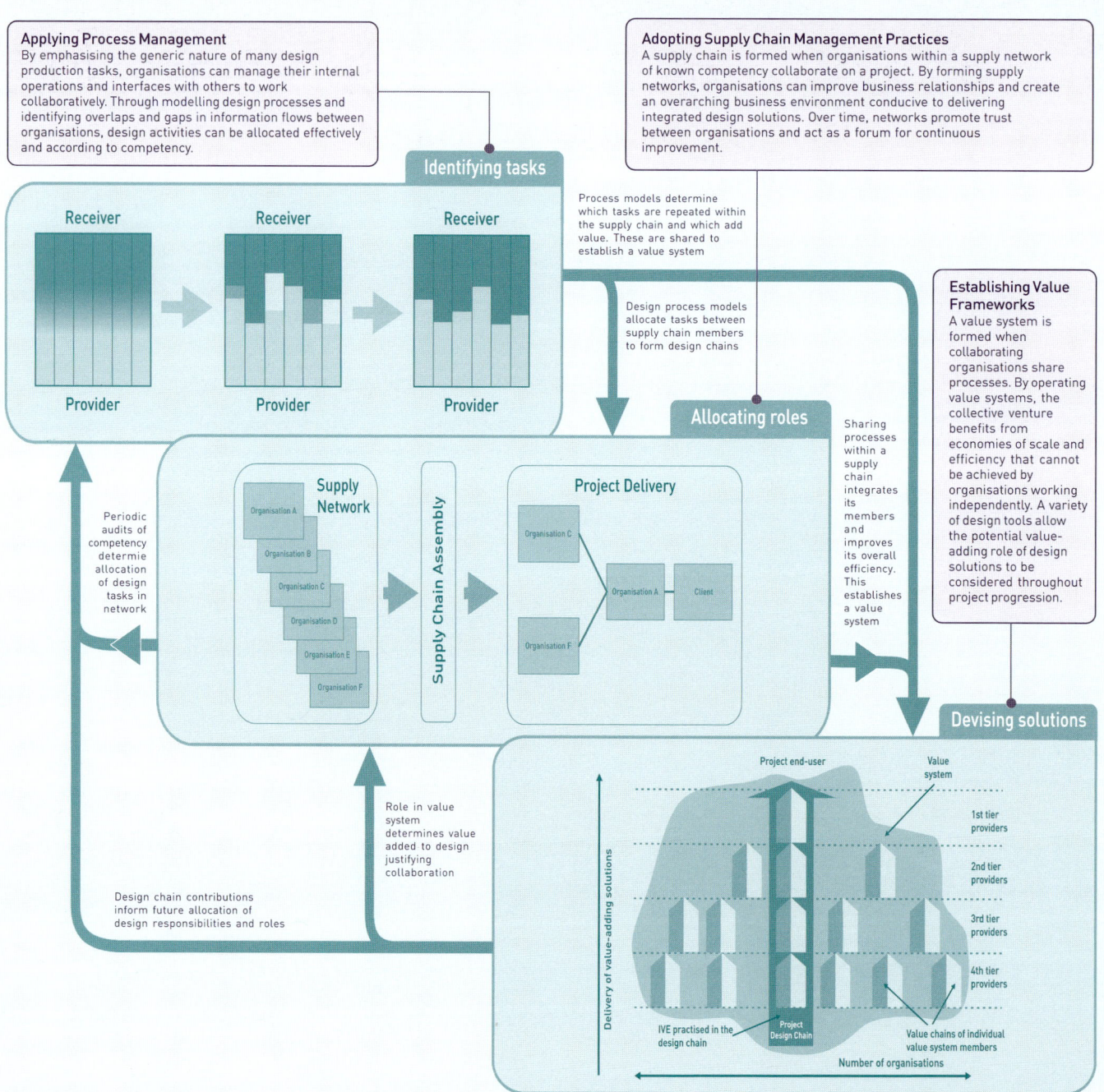

Identifying tasks

Process models determine which tasks are repeated within the supply chain and which add value. These are shared to establish a value system

Design process models allocate tasks between supply chain members to form design chains

Allocating roles

Establishing Value Frameworks
A value system is formed when collaborating organisations share processes. By operating value systems, the collective venture benefits from economies of scale and efficiency that cannot be achieved by organisations working independently. A variety of design tools allow the potential value-adding role of design solutions to be considered throughout project progression.

Sharing processes within a supply chain integrates its members and improves its overall efficiency. This establishes a value system

Periodic audits of competency determie allocation of design tasks in network

Devising solutions

Role in value system determines value added to design justifying collaboration

Design chain contributions inform future allocation of design responsibilities and roles

Part 5: The ICD Practices

5.1   Introduction

This part presents the practices that an organisation may use to deploy its ICD approach to design management. They are classified by their type and the domain to which they are applied, giving six categories:

1. Business Strategies
2. Business Tactics
3. Business Operations
4. Project Strategies
5. Project Tactics
6. Project Operations

> The ICD terminology is explained in the Glossary at the back of this handbook.

You should use your maturity assessments from Section 3.5 (Tables 3.2, 3.3 and 3.4) and the guidance provided in Part 4 (Sections 4.2 through 4.6) to determine which practices are relevant to your needs, given the role you and/or your organisation plays in the *design chain* (as a *provider* or as a *receiver*) and whether you are concerned with managing activities in the *business domain* or the *project domain*. Note that, within each category, the numbering provided is for reference purposes only, and has no other significance.

> For further supporting information on these practices please visit the design chains website at **www.designchains.com.**

5.2 Practice Format

To aid their comparison, each practice is presented in a standard format comprising seven parts (Table 5.1). Where appropriate, additional guidance notes highlight issues of specific concern to providers and receivers.

Section	Description
Purpose	Describes the intended application of the practice.
Summary	A short overview of the practice, outlining its key elements and its method of use.
Provider Benefits	Lists the main benefits of the practice that are relevant to providers.
Receiver Benefits	Lists the main benefits of the practice that are relevant to receivers.
Outline Procedure	Presents a summary of the key stages of the practice. All strategies share the same basic procedure according to whether they apply to the business or project domains. Similarly, some tactics share a common procedural format.
Requirements and Resources	Describes the types of people and the infrastructure required to support the practice, if it is to be most effective. This includes related ICD or other provisions that must be in place before the practice can be applied.
Related Practices	This tabular section identifies other practices, which are cross-referenced in the practice being described.

Table 5.1:
Structure of ICD practice descriptions

BS1

Planning Design Process Management

Provider benefits

- By introducing a process view into the supply network, both provider and receiver organisations will be preparing to work in an integrated and collaborative manner by enhancing the division of responsibilities.

- Providers have the opportunity to compare and harmonise their process and working practices with other providers and receivers within a supply network or a design chain.

Receiver benefits

- If receivers monitor their internal processes they can plan improvements to the way in which they collaborate with providers through an alignment of outcomes for each work activity and the establishment of coherent stage gates.

- Process management improvements can address specific project issues or specific aspects of project management activity. This targeted approach should help organisations focus on generating a real understanding of critical interfaces, avoiding the inherent risks in attempting to manage concurrent improvements in processes.

Purpose

Organisations apply process management to the business in order to capture, communicate and maintain a generic representation of business and project design processes. This helps them to understand the tasks that must be performed and to identify who in the supply network is best placed to take responsibility. Process management is maintained in the business domain, where it is used to guide project activities. By developing a shared understanding of the generic project process, supply network organisations can establish collaboration within the business domain that will lead to the simplification of their interaction when they work together on projects, increasing efficiency and effectiveness.

Summary

Planning an organisation's design process management practice is concerned with making sure that each ICD organisation is prepared to adopt and maintain a coherent view of their business and project operations as processes which can be co-ordinated. This practice is concerned with planning the various aspects of design process management, so that it is aligned with the organisation's strategic direction, and effective communication throughout the supply network. By establishing a framework of language and process, organisations can develop alignment with others in the supply network to reconcile different perspectives in terms of risk, cost planning, and alignment of design deliverables.

Organisations should plan design process management in a way that develops an internal understanding of business and project processes. This involves reviewing existing methods for representing generic processes, and adopting an appropriate method in the business domain. This will allow supply network members to align themselves internally through their improved understanding of their internal processes and information flow, and externally by understanding the inter-organisational flows of information. When adopting a process view, ICD organisations should review the maturity of their experience of operating within a design chain. This will enable them to make tactical decisions relating to the type of process modelling to be undertaken and to what level of detail.

Outline Procedure

1. Where are we now?
2. Where do we want to be?
3. How do we get there?
4. How well are we doing?

Requirements and Resources

- The ICD principle of Adopting Process Management must be established within the business. This is required to ensure that processes can be allocated between business partners (on the basis of which party is best placed in terms of competency and risk management) and links between processes maintained, even when these cross the interfaces between organisations.
- The strategy for planning design process management should be supported by the necessary tactics and operations in the business and project domain.

Related Practices

- ☐ **BS1** Planning Design Process Management
- ☐ **BS2** Planning Supply Chain Management Business Practice
- ☐ **BS3** Planning the Implementation of Integral Value Engineering across the Business

- ☑ **BT1** Applying Process Management in the Business
- ☐ **BT2** Aligning Supply Networks
- ☐ **BT3** Applying Supply Chain Management in the Business
- ☐ **BT4** Applying Integral Value Engineering in the Business
- ☐ **BT5** Conducting a Value Survey
- ☐ **BT6** Performing an ADePT Review

- ☑ **BO1** Modelling Business Design Processes
- ☐ **BO2** Auditing the Supply Network
- ☐ **BO3** Auditing the Supply Network for Technical Competence
- ☐ **BO4** Gathering Value-adding Feedback from Projects

- ☑ **PS1** Planning Project Design Management
- ☐ **PS2** Planning Supply Chain Management Project Practice
- ☐ **PS3** Planning Integral Value Engineering Project Practice

- ☐ **PT1** Applying Design Management Practices
- ☐ **PT2** Applying Supply Chain Management to a Project
- ☐ **PT3** Selecting Supply Chain Members at the Project Level
- ☐ **PT4** Implementing Integral Value Engineering on a Project

- ☐ **PO1** Applying ADePT to Design Management
- ☐ **PO2** Applying DePlan to Design Management
- ☐ **PO3** Modelling Project Design Processes
- ☐ **PO4** Assembling Supply Chains
- ☐ **PO5** Applying Value-adding Tools to Design Problems

Where are we now?

This stage is concerned with gathering information on current business operations and practices to inform your strategic plans.

Review Assessment of ICD Principles in your Organisation
Your strategy should be developed in relation to the current status of ICD principles within your organisation. Review the findings of the assessment you performed in Table 3.2 of Section 3.5 to determine this. You should aim for balance and consistency across all of the characteristics of process management, ensuring that the supply network develops its understanding of processes together.

Review Organisation Internally
Your organisation's working practices should be reviewed to summarise how it currently practices process management, including the resources used to support its functions. Your organisation will need to determine its understanding of:

- your general processes, including your experience of adopting a process view as a business and as part of a design chain;
- the use of a high-level process framework, such as the RIBA *Plan of Work*;
- low-level design process maps, often built up from well-established work-breakdown structures, showing how you organisation undertakes design activity in detail;
- your current ability to manage processes, identifying the resources needed to adopt process management; and assigning roles and responsibilities to key individuals;
- how you currently plan your processes, in both the project and business domains; and
- the potential barriers to change and to improvement of your processes.

Review Business Partners
The practice of process management across all supply network members should be reviewed to determine how your organisation currently works with other organisations. This will help you develop strategies that build upon these existing ways of working to achieve greater collaboration and integration in the design process. Carrying out combined maturity assessments with key partners may assist collaboration. While full alignment of business and delivery processes may not be possible, it should be possible to create an understanding of the key stage-gates within a supply network.

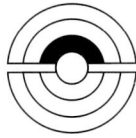

Where do we want to be?

This stage is concerned with setting appropriate goals in light of the current status of process management within your organisation and the time period over which it will be developed.

Review Internal Business Goals

Each ICD organisation's design process management practice must be aligned to both its business and its project domain activities. When considering your organisation's operations, providers and receivers must be sure that the business goals are met through the planned deployment of an appropriate approach to design process management. Once generic processes are established, they can be amended and/or developed to suit any change in business or project strategies. When operating within a supply network or design chain, ICD organisations must be able to align themselves through the mutual understanding of both internal and external processes.

Review External Business Goals

Your organisation's practice of process management should be managed in a way that will build upon your relationships with other organisations. Appropriate goals for developing working practices to achieve this should be set. This essential element of all business strategies is particularly important for process management deployment, given the opportunities it creates for ICD. It may be possible to set up overriding processes for all members of a supply network, where membership of the network is conditional upon accepting a common process view (e.g. advanced SCM frameworks, such as that of BAA). However, for many businesses the investment in developing a single process framework would be prohibitive. Different process frameworks will be acceptable, provided that the interfaces at critical stage-gates are aligned in terms of risk, cost and design.

Set Strategic Targets

The implementation of design process management needs to be planned alongside other strategic developments and take into account all parts of an organisation's process management approach (see, Planning Project Design Management (PS1) and Applying Process Management in the Business (BT1)). Design process management requires the identification of a suitable process framework. ICD organisations can work with a number of existing process frameworks, and these can be adopted to suit the needs of each individual organisation.

There are several process frameworks in common use, the most established being the RIBA *Plan of Work*. This has proved to be too restrictive for many organisations, leading to the development of specific process maps for particular clients, disciplines and projects. The Process Protocol (see Figure 2.8) is a collaborative attempt to develop a generic process framework. It is likely that organisation and supply networks will seek to develop bespoke process management frameworks aligned to particular business or project needs.

Receivers

- Receivers should be aware of the provider strategy to be sure that they fully understand their design process management practice and remain aligned with supply network members.

How do we get there?

This stage is concerned with planning a logical development of your working methods to advance your process management.

Define Strategic Scope
Define the required scope of change by comparing your organisation's current process management practice with those required. This will determine the extent of the change in your organisation's working practice that your strategy must realise. It is more productive to set realistic goals for incremental improvement or staged adoption of new principles, rather than seeking to achieve complete alignment of process management practices in a short time-scale.

Select Deployment Method
Depending on the scope of the change required, you may choose to implement your strategy as a series of step-changes in the development of your organisation's working practices. Alternatively, if your strategy requires the adoption of a new or different way of working, you should broaden its deployment to include the training and education of all employees to manage the change. Consequently, you may need a more hands-on method of deploying your strategy, establishing change processes for internal employees and external partnerships. Consider whether the process management needs of a section of your supply network could be developed as a pilot or demonstration project. It is likely that, in most cases, you will deploy your strategy using a combination of methods.

Plan Deployment
The adoption of a process view within an organisation's business domain activities may require a significant change to the way in which it operates. Therefore, it is important to involve key individuals to support and drive the implementation plan forward. When planning strategy implementation, organisations must make provisions for revised strategic plans that may arise as a result of changing business needs. Approaches to planning strategy deployment include:

- setting short-term goals (e.g. 100% deployment internally by....) and medium-term objectives (e.g. top ten supply partners integrated by .....);
- embracing those parties who are most nearly aligned as leaders; and
- encouraging those individuals and organisations who find the change difficult.

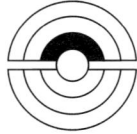

How well are we doing?

This stage is concerned with monitoring the effect your strategy is having, so that you learn from its deployment and improve your business.

Determine Feedback Required
Your strategy should be adaptable during deployment to allow for any obstacles encountered, capturing both positive and negative feedback. Identify the key performance measures, especially those associated with the strategic targets.

Establish Feedback Mechanisms
The maturity of process management competencies should be kept under reveiw. The outcomes should be monitored and the strategy revised if it is not having the intended effect and/or to take account of any changes that may have occurred in the business environment. It is important that analysis of feedback accurately identifies cause-and-effect relationships.

Gather and Respond to Feedback
Once process management is established and underway, it is important that it remains appropriate to the operating environment. In a similar manner to responding to change when establishing process management practice, this continuous review helps the organisation determine if its emerging process management practice remains compatible with the latest operating needs of the business. For example, a mechanism to record the understanding (or lack of understanding) of process frameworks within a supply network is required to inform the alignment of processes between network members.

BS2

Planning Supply Chain Management Business Practice

Provider benefits

- Providers develop an understanding of SCM and the intent of a receiver who is adopting SCM.

- Planning for SCM (by the receiver) allows the provider to determine whether it is relevant to the provider's business and how it should engage with other organisations who are adopting SCM.

Purpose

To enable the development of a plan to adopt SCM practices within the business domain. This practice outlines issues that influence how an organisation plans and implements SCM. These include confirming the appropriateness of SCM to the organisation, communicating this need, and the actions necessary to allow it to take place.

Summary

ICD practice does not prescribe a given way to introduce SCM (and the training it requires) into an organisation. What follows are the different approaches that organisations (typically receivers) can take and how other organisations can implement SCM or respond to SCM being implemented by others.

The scope of SCM is broad, extending out from an organisation to encompass its relationships with other companies and its customers. For the individual company, integrating SCM into business practices can require a step-change in the company's values, attitudes and practices. This practice describes a variety of ways that this can be achieved.

Receiver benefits

- Receivers develop an understanding of SCM.

- Planning for SCM gives a clear understanding of the need to adopt SCM practices.

- Planning gives an early understanding of any potential barriers to implementation and thus the opportunity to develop ways of overcoming them.

- An appreciation of the importance of sending consistent strategic signals to staff within the organisation and to other design chain members.

- An awareness of the range of mechanisms available for the company to communicate the need for a change in values both internally to staff and externally to other members of the design chain.

Outline Procedure

SCM can be used to inform business strategy through a four-stage process:
1. Where are we now?
2. Where do we want to be?
3. How do we get there?
4. How well are we doing?

Requirements and Resources

- No specific requirements exist for this practice, as it is the precursor for other SCM practices. However, process management is an enabler. Where used in conjunction with Auditing the Supply Network (BO2), it represents an opportunity to review the organisation's objectives but requires the supply network to have been audited beforehand.
- The individuals who will provide input to this practice should be drawn from a variety of areas within the business rather than a single function. It is recommended that the process involves senior management to provide the necessary strategic input, together with project personnel to ensure that the style of SCM introduced is relevant to existing operational activities.

Related Practices

- ☐ **BS1** Planning Design Process Management
- ☐ **BS2** Planning Supply Chain Management Business Practice
- ☐ **BS3** Planning the Implementation of Integral Value Engineering across the Business

- ☐ **BT1** Applying Process Management in the Business
- ☐ **BT2** Aligning Supply Networks
- ☑ **BT3** Applying Supply Chain Management in the Business
- ☐ **BT4** Applying Integral Value Engineering in the Business
- ☐ **BT5** Conducting a Value Survey
- ☐ **BT6** Performing an ADePT Review

- ☑ **BO1** Modelling Business Design Processes
- ☑ **BO2** Auditing the Supply Network
- ☐ **BO3** Auditing the Supply Network for Technical Competence
- ☐ **BO4** Gathering Value-adding Feedback from Projects

- ☐ **PS1** Planning Project Design Management
- ☐ **PS2** Planning Supply Chain Management Project Practice
- ☐ **PS3** Planning Integral Value Engineering Project Practice

- ☐ **PT1** Applying Design Management Practices
- ☐ **PT2** Applying Supply Chain Management to a Project
- ☐ **PT3** Selecting Supply Chain Members at the Project Level
- ☐ **PT4 Implementing Integral ValueEngineering on a Project**

- ☐ **PO1** Applying ADePT to Design Management
- ☐ **PO2** Applying DePlan to Design Management
- ☐ **PO3** Modelling Project Design Processes
- ☑ **PO4** Assembling Supply Chains
- ☐ **PO5** Applying Value-adding Tools to Design Problems

Where are we now?

This stage is concerned with gathering information to inform your strategic plans.

Review Assessment of ICD Principles in your Organisation
Your strategy should be developed with regard to the current status of ICD principles within your organisation. Review the findings of the assessment you performed in Table 3.3 to determine this.

Are Supply Networks Relevant to your Business?
The ICD approach of using supply networks to support an organisation's adoption of SCM practices is only one way of managing organisational relationships. Each organisation must assess the relevance of a supply network to its operational needs. This requires a review of the nature of an organisation's business, which might typically be done through a workshop, attended by representatives of all areas of the business. The products and services comprising an organisation's business can be mapped against a six-cell matrix plotting frequency of demand against the type of product or service exchanged[1]. Within an ICD approach, the management of business practice is focused primarily on the exchange of design information so that, for example, a heating and ventilation design specialist may see themselves as repeatedly selling a customised design service to another organisation.

If a workshop is held, it should challenge existing assumptions about the organisation's business:

- Does the organisation provide a product or a service? How does it provide value to the end-user?

- Which markets does it serve? Who are its receivers? Does the organisation sell to repeat customers? The smaller the number of customers, the greater the customisation of the product or the services, and the greater likelihood that SCM will be advantageous.

- From which markets does the organisation procure? Who are its providers? Do they provide customised or commodity products and/or services? Are they strategically important?

		Characteristics of subject of exchange (goods and services)		
		Non-specific (commodities)	Mixed	Idiosyncratic (bespoke)
Frequency	Occasional	Purchasing standard equipment	Purchasing customised equipment or services	Construction of a building
	Recurrent	Puchasing standard material	Purchasing customised material or services	The site specific transfer of intermediate products or services across sucessive stages

Reviewing an Organisation's Purchasing History
A strategic purchasing review looks at the historic relationship between the receiver and its providers. It can be used to identify patterns in existing purchasing arrangements and whether this might indicate if any informal supply networks already exist.

Recognising the Need for Internal Alignment
To co-ordinate with other members in a design chain, each member must first ensure that different departments and specialisations within its own company have a common understanding. Often parts of a business, by seeking to improve their own performance, form functional silos that damage the firm's overall performance within which different working attitudes can evolve. These internal barriers need to be broken down.

Does an Organisational Consensus exist for Adopting SCM?
Because SCM is a 'philosophy' that links together a range of practices, rather than being a practice in its own right, that philosophy needs to be shared for it's successful adoption. As part of the initial review, there is a need to assess whether any consensus exists within the organisation for adopting SCM.

Continued overleaf

[1]Williamson, O. E. (1979) Transaction-Cost Economics: The Governance of Contractual Relationships, *Journal of Law and Economics*, 22 (2) October pp.233-261

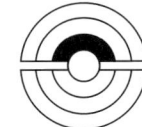

Continued from previous page

What Other Management Initiatives are being Undertaken in the Organisation?
ICD practices reflect a process perspective that is shared by other management concepts, such as quality assurance or business process reengineering. Other initiatives or activities within the organisation may already address issues relevant to SCM and so may be co-opted into ICD practice, such as the mapping of existing processes or procedures in quality assurance manuals. Organisations need to assess the nature and extent of existing procedures that may be relevant to the adoption of SCM and whether these differ between business or operational units.

Providers

- If individual providers are finding it difficult to co-ordinate with other providers when they serve a common receiver, they may identify the need for a supply network through their assessment. Additionally, providers may together instigate the formation of supply networks to serve a range of receivers.

Receivers

- The members of the supply network need to be clear on how they think SCM will be used within its member organisations and how it will relate to the individual business plans of each organisation in the network. For example, SCM is particularly suited to organisations that seek to gain more control over the selection of members for supply chains.

Where do we want to be?

This stage is concerned with setting appropriate goals, in light of the current status of adopting SCM practices within your organisation and the time period over which it will be developed.

Set Internal Business Goals
ICD builds on the foundation of process management to provide a consistent and coherent approach to managing the delivery of projects. To achieve this through SCM practices, it is necessary to set internal goals (as outlined below).

Align Organisations Internally
The goal of aligning the different parts of an organisation may involve:
- providing a shared view of the organisation and its relationship with its providers and receivers through the strengthening of the organisational values and SCM culture; and
- providing explicit controls over how people should work and behave that are consistent across the organisation and across projects through the existence of 'core' procedures.

Build an Organisational Consensus
The goal of building a consensus within an organisation reflects the need not only of communicating the business case but proving that it can be delivered to ensure the psychological 'buy-in' necessary for people to move voluntarily to working in more integrated and collaborative ways.

Set External Business Goals
A business attempting to rationalise their supply chain should identify the different types of business relationships they need to engage in with providers and which organisations they would typically involve. For each of the market segments that a receiver procures from or a provider sells to, it can allocate the product and/ or service it procures or provides to one of the four cells in the figure below.

This will enable any organisation to establish which products or services represent areas where SCM is most applicable.

Set Strategic Targets
Business goals both internally and externally may result in targets that vary from the general (building an organisational consensus to support SCM) to the more technical development of specific ways of working. Detailed purchasing reviews which track 'where the money goes' can help strategic planning in the setting of targets for numbers of suppliers to be used or setting limits to the proportion of workload that a particular provider may represent.

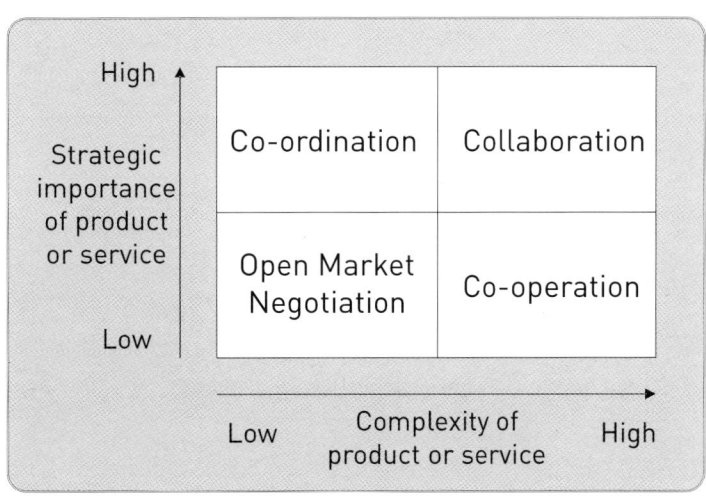

Receivers

- An SCM audit (see Auditing the Supply Network (BO2)) can form part of an organisation's SCM development strategy. It can also help the business development plan to ensure that it responds to changes in the organisation's operating environment during SCM implementation.

- SCM has a clear role to play in helping to frame an organisation's strategy, and to select providers for admission into new and established supply networks (see Auditing the Supply Network (BO2)). However, a strategic review of purchasing will be required before SCM can extend into the business domain (see Applying Supply Chain Management in the Business (BT3)).

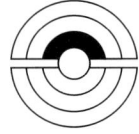

How do we get there?

This stage is concerned with planning a logical development of your working practice to advance your practice of ICD.

Strategic Reviews of Purchasing

Strategic reviews of purchasing provide a mechanism to achieve a smooth and stabilised supply network membership, by ensuring that work is distributed appropriately among its members and the network is not dominated by relationships that are too strong between a receiver and any particular provider within the network. This avoids the risk of companies placing 'all their eggs in one basket', and ensure that all members of the network feel they are being treated fairly.

Encouraging Internal Alignment

Through the Development of Core Values
The large, varied range of projects an organisation may undertake can make it difficult to control behaviour through formal practices. However, organisations can use corporate values as an indirect means of control. Middle management (in both the project and business domains) can then translate these values into work practices. In time, designers will absorb the values, via the practices which embody them, and make the company's values their own.

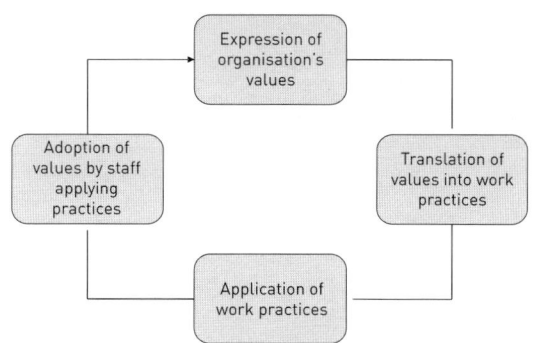

Through the Development of Adaptable Procedures
Senior management must establish a balance between the need to define working practices rigidly and the need to allow sufficient flexibility to accommodate variations between projects. Failure to achieve this will undermine the credibility of the formal processes and will encourage ad hoc arrangements to evolve, resulting in a range of different ways of working across the organisation. Therefore, organisations therefore need to distinguish between their preferred (most likely new) ways of working that are promoted as guidance representing how they wish to work with the more static, formal procedures that define the core, fundamental practices that it must always perform.

Through the Development of Specific Ways of Working
The ICD approach is particularly suited to the development of core business and to project activities that include generic processes. Design process models (see Modelling Business Design Processes (BO1)) can help organisations understand the activities they perform and which of these are generic and which vary significantly from project to project. Specific ways of working may be very technical in nature and are defined by a design chain member to structure its function within the chain and prescribe how certain processes are to be performed and, in doing so, embody its corporate values. They represent how that organisation wishes all projects to progress and that ideal is communicated both internally and externally.

Building Organisational Consensus

Because SCM is a 'philosophy' that links together a range of practices, rather than being a practice in its own right, it needs to be introduced by building a consensus in both the business and project domains. The business case for adopting SCM practices can be communicated in a variety of ways, including newsletters or presentations. Senior management, however, must first decide that supply networks are required and exercises, such as the assessment of need, strategic purchasing reviews, project reviews (see Assembling Supply Chains (PO4)) and the feedback from traditional marketing activities, can all supply information to inform strategic decision-making. SCM as a solution can be sold to organisation members most effectively if:

It delivers clear and tangible early benefits
A major difficulty in changing established practices is proving the benefit of a different way of working. The relatively long life of projects and the industry's inability to measure benefits until projects have concluded introduce long delays to the gathering of evidence in support of different ways of working. In particular, the financial success of a project (often used to determine the success of any new methods) is not known until a Final Account has been agreed. Consequently, individuals tend to view new initiatives as not focusing on the

Continued on next page

Continued from previous page

'base issues' of finance or money, hindering their adoption.

It is shown as already being applied
Many of the collected practices in SCM are applied in isolation by organisations. Experiences on other projects or in other areas of the business can be used as case studies or demonstration projects of what can be achieved. SCM can then be presented as merely broadening the application of those practices.

It is shown to be inevitable
The likeliness of an outcome can be increased through the appointment of particular steering groups and specialist personnel, or by the recommendations of research and development programmes. Using examples from business and project processes and procedures to acknowledge desirable behaviour is also a mechanism for transmitting values within a company.

With design, the direct benefits of new working methods are often hard to measure. Financial figures tend to be measured on a whole-project basis, preventing the accurate assessment of individual initiatives in isolation. Further, anecdotal evidence (lack of day-to day conflict, for example) is difficult to evaluate in any comparative way. To address this, the introduction of SCM within an ICD approach should look to justify itself as soon as possible. This can be achieved by the following:

- *Rationalising existing approved lists of providers*. The lists may have become outdated and might include organisations that may have only worked once for a receiver (typically upon a client's request). By reviewing approved lists, provider numbers can be trimmed down to provide something that all can see is useful.

- *Project reviews*. Frequently, new practices are applied within projects without reference to previous project experience. It can be avoided by implementing project reviews so that benefit can be documented to inform (and justify) the repetition of new practices on further projects. As this continues to happen from project to project, new practices become established and become part of 'the way things are done'. At this mature stage, it is likely that they will not be addressed specifically in project reviews.

- *Use of existing management controls to provide benchmarks*. Provisional benchmarks can be created quickly by analysing basic data that are collected in the normal course of business. Project data are a little harder to come by but may be found from existing sources (such as time sheets or logs of technical queries) which can then be investigated as potential indicators of poorly managed areas of design integration.

Building on Existing Initiatives
Viewing projects as supply chains provides a context in which new working practices can be introduced and built upon. These can be used to bind network members together to provide strategic direction. Examples include the following:

- Business Process Reengineering (BPR), which can be used to develop a process view of a business and its projects.

- Quality assurance can help in two ways, in the first place, by documenting current processes, it can communicate messages about the company's values. Secondly, because it seeks to bring everything a company does up to a consistent standard, the messages being communicated are also likely to be consistent.

- The EFQM Business Excellence Model can help an organisation address a series of business domain issues and, in the process, prepare the ground for supply networks to be established.

- Investors in People can act as an enabler by assisting organisations to establish a good internal communication infrastructure that can then be used to promote cultural change within the organisation.

- Internal accounting practices are something that can be changed to encourage preferred forms of behaviour, such as changing internal budgeting arrangements to focus on the out-turn cost rather than on tender prices.

By 'pushing on open doors', the introduction of new working practices can be eased by avoiding 'initiative overload' and the corresponding drift back to more comfortable ways of working. Many initiatives apply to both receivers and providers. In all cases, they can be refined to ensure a common message is given as to an organisation's intended way of working.

Continued overleaf

Continued from previous page

Providers

- SCM places providers in a strong position to sell the need for technical or organisational change up the design chain because it provides a mechanism to exert greater influence over the receiver and requires receivers to be receptive to input from providers. However, if not managed carefully, a provider might only react to external pressures (such as from the receiver), falling back into the traditional, reactive approach of much of the construction industry.

- A receiver's use of a small number of core procedures allows its behaviour to be anticipated because it is more stable and, thus, more predictable.

Receivers

- Receivers need to be aware of how their intentions will perceived by providers and by their own staff to avoid the cynical view that proposed changes are for the benefit of providers but at the expense of receivers rather than being seen as a 'win-win' outcome. With time, this problem disappears because receivers will have been able to demonstrate their intentions by their actions on projects. However, in the early stages of ICD, confused signals can hinder relationships within receiver and provider organisations.

- In proposing changes to working practices, providers do not necessarily need to demonstrate the benefits of the changes to receivers. However, to ensure a smooth relationship and support from other providers in the design chain, the advantages of such changes on the provider's existing ways of working need to be spelt out. Changes to working practices will clearly only occur where organisations and individuals perceive it to be in their mutual interest.

How well are we doing?

This stage is concerned with monitoring the effect your strategy is having, so that you can take corrective action, if required. In all four of the following areas, there is as much that can be learnt from what has not happened as from what happened. Failures in SCM, (e.g. the unwillingness of operational units in an organisation to develop a common supply network) are made evident on projects, but may often reflect problems in the business domain.

Reviewing the Relevance of Supply Networks to your Business

This activity is essentially a repetition of the initial assessment as to the relevance of supply networks. The process must involve senior management (to provide the necessary strategic input), as well as key operations personnel to ensure the relevance of a supply network to the company's operational activities. With increasing SCM maturity, it may be appropriate to develop the assessment process from subjective to evidence-based assessments (evidenced through documentation, for example), thus providing a firm baseline from which to seek improvement.

Strategic Reviews of Purchasing

The review of the historical purchasing profile of the organisation mapped against the clear targets set by the strategy can provide an objective measure of progress in the adoption of some aspects of SCM practice. Following the review, decisions can be made as to whether the existing targets are appropriate.

How Internally Aligned is your Organisation?

The degree to which internal barriers have been removed within an organisation is a difficult one to measure. Techniques exist to assess the degree to which values are shared but these tend to be of limited use to industry practitioners. Of more use are anecdotal indicators, such as where problems arise in internal communication and the subjective impressions of personnel across an organisation, which may be captured through a review workshop or the use of questionnaires for example.

Does an Organisational Consensus exist for Adopting SCM?

The extent of 'buy-in' to the philosophy of SCM needs to be continuously monitored. Management initiatives are liable to fall in and out of fashion, but the nature of SCM as a concept that communicates the extent of the interdependence that exists between organisations in the supply chain, requires that the underlying philosophy is nurtured and maintained. This SCM philosophy should also be communicated internally to new staff and externally to other organisations.

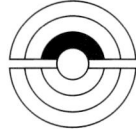

BS3

Planning the Implementation of Integral Value Engineering across the Business

Provider benefits

- When a receiver co-ordinates the introduction of IVE into a supply network (or a value system built upon it) there is an opportunity for collaboration between providers and other organisations which can then be marketed to clients (i.e. receivers) as a common competency.

Receiver benefits

- By aligning IVE practices, receivers and providers can agree, independently of any project, the basis for their value-related collaboration on projects. Receivers are in a good position to do this as they can use their existing knowledge of providers' competencies provided by periodically Auditing the Supply Network (BO2).

- By taking a strategic perspective when planning IVE, a receiver can solicit feedback and adjust its plans (or those of the value system that it is co-ordinating) accordingly. This will ensure it remains responsive to an evolving operating environment, including the evolving roles of providers as their SCM competencies and maturity develops.

Purpose

This strategy will help an organisation plan its approach to Integral Value Engineering (IVE) alongside its existing practices and long-term objectives. It is adopted in the business domain to help satisfy client and stakeholder project needs, demonstrate value for money, and to ensure consideration of value throughout projects.

Summary

Planning IVE is concerned with aligning relevant elements of an organisation's knowledge, resources, culture, working methods and tools. It involves planning to ensure that these elements are compatible with the organisation's strategic direction and ongoing partner relationships. IVE practices are maintained in the business domain so that they can be used with consistency from one project to another.

Organisations should plan the development, based on the assessment of their IVE practice in Section 3.5, as a series of logical steps. Initial steps should establish those elements of their ICD approach concerned with establishing IVE within the organisation's corporate values. If the organisation is a member of a value system, co-ordinating these steps across all value system members will allow a coherent approach to IVE to be practised across the value system irrespective of the individual organisations involved in specific projects. This consistency in approach to IVE throughout its development can be sold to value system clients.

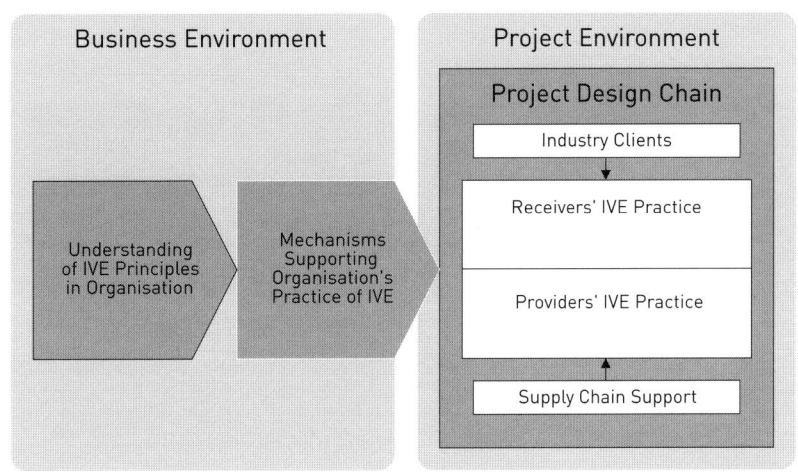

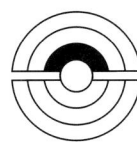

Outline Procedure

1. Where are we now?
2. Where do we want to be?
3. How do we get there?
4. How well are we doing?

Requirements and Resources

- A strategy to establish IVE business practice is only effective when supported by tactics and operations in the business domain and the project domain. Each tactic and operation must be deployed in line with the business strategy.

Related Practices

- ☐ **BS1** Planning Design Process Management
- ☐ **BS2** Planning Supply Chain Management Business Practice
- ☐ **BS3** Planning the Implementation of Integral Value Engineering across the Business

- ☐ **BT1** Applying Process Management in the Business
- ☐ **BT2** Aligning Supply Networks
- ☐ **BT3** Applying Supply Chain Management in the Business
- ☑ **BT4** Applying Integral Value Engineering in the Business
- ☐ **BT5** Conducting a Value Survey
- ☐ **BT6** Performing an ADePT Review

- ☑ **BO1** Modelling Business Design Processes
- ☑ **BO2** Auditing the Supply Network
- ☐ **BO3** Auditing the Supply Network for Technical Competence
- ☐ **BO4** Gathering Value-adding Feedback from Projects

- ☐ **PS1** Planning Project Design Management
- ☐ **PS2** Planning Supply Chain Management Project Practice
- ☑ **PS3** Planning Integral Value Engineering Project Practice

- ☐ **PT1** Applying Design Management Practices
- ☐ **PT2** Applying Supply Chain Management to a Project
- ☐ **PT3** Selecting Supply Chain Members at the Project Level
- ☐ **PT4** Implementing Integral Value Engineering on a Project

- ☐ **PO1** Applying ADePT to Design Management
- ☐ **PO2** Applying DePlan to Design Management
- ☐ **PO3** Modelling Project Design Processes
- ☐ **PO4** Assembling Supply Chains
- ☐ **PO5** Applying Value-adding Tools to Design Problems

Where are we now?

This stage is concerned with gathering information on your current business operations and working practices to inform your strategic planning.

Review Adoption of ICD Principles in Organisation
Your strategy should be developed with regard to the current status of ICD principles within your organisation. Review the findings of the assessment you performed in Table 3.4 of Section 3.5 to determine this.

Review Organisation Internally
Your organisation's working practices should be reviewed to summarise how it currently practices IVE, including the various IVE resources used to support its function. Review your organisation's:

- existing understanding of value;
- establishment and communication of client needs as project team values;
- use of value management to understand and synthesise these project values among project stakeholders;
- barriers to the address and delivery of value; and
- any other issues relevant to your organisation's ability to deliver value to clients.

This will allow you to determine the manner by which your organisation currently works with other value system members so that you can develop strategies to build upon current working practices.

Review Business Partners
If your organisation operates within a value system or supply network, then the above internal review should be repeated on behalf of all value system members. Instead of addressing IVE within a single organisation (including the maintenance and use of IVE resources, such as a value-adding toolbox), the combined practice of IVE across all value system members should be reviewed. This will determine how your organisation currently works with other value system members so that you can develop strategies to build upon current working practices.

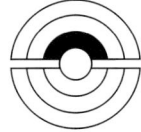

Where do we want to be?

This stage is concerned with setting appropriate goals in light of the current status of value frameworks within your organisation and the time period over which it will be developed.

Review Internal Business Goals

IVE must compliment other business domain activities. The current key issues for the organisation (such as its position in its market place, employee attitudes towards change and new technology) must therefore be identified and planned, learning wherever possible from past experience and knowledgeable individuals.

Review External Business Goals

Your organisation's practice of IVE should be managed to build relationships upon it with other organisations in its supply network or value system. Appropriate goals for developing the manner by which it works with others should therefore be set. This essential element of all business strategies is particularly important for IVE development, given the opportunities it creates for collaborative design and the integration of business partner organisations to form a value system.

Set Strategic Targets

The implementation of IVE needs to be planned alongside other strategic developments; in particular, it needs to take account of other parts of an organisation's, or a value system's, ICD approach: see Applying Integral Value Engineering in the Business (BT4) and Planning Integral Value Engineering Project Practice (PS3).

Providers

- Providers should contribute their experiences of change to the receiver establishing IVE practice on behalf of the value system. This ensures that the deployment strategy can be aligned with the experience of others in the value system and, thereby, reducing resistance to change.

Receivers

- By co-ordinating your strategy with the ongoing business development of supporting providers, a consistent understanding of IVE practice will be maintained throughout a value system, irrespective of the current stage of strategy implementation. This consistency can be marketed to clients as a differentiating feature of the organisation or value system.

How do we get there?

This stage is concerned with planning a logical development of your working practice to advance your value frameworks.

Define Strategic Scope

Define the required scope of change by comparing your organisation's current IVE practice with those required. This will determine the extent of the change in your organisation's working practices that your strategy must realise.

Select Deployment Method

Depending on the scope of the change required, you may choose to implement your strategy as a series of step-changes in the development of your organisation's working practice. Alternatively, if your strategy requires the adoption of a new or different ways of working, you should broaden its deployment to include the training and education of all organisation employees (see figure below). This may help manage the change in the values of your organisation required to buy-in to different ways of working. Consequentially, you may adopt a more hands-on method of deploying your strategy. It is likely that, in most cases, you will deploy your strategy using a combination of methods.

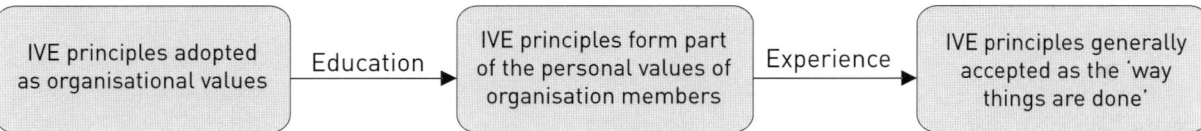

By making deployment a continual process (i.e. by not stopping and starting the change process), the buy-in of all parties can be sustained, ensuring a logical evolution in the understanding of IVE business practice. This creates opportunities for each organisation to gradually align their IVE practice both internally and with other value system members, while also ensuring that they become fully competent at each stage before moving to the next.

Plan Deployment

Because integrating IVE into an organisation's business domain activities can require extensive change to the way that organisation works, it may take a long time. Key personnel (sufficiently senior to drive the strategy forward) will be needed to sustain planned implementation. 'Change champions' may also need to be appointed and distributed throughout the organisation. Such champions can also help identify changes in the business environment that might necessitate revisions to strategic plans for IVE development.

Your organisation should also plan the training of IT staff and management in the provision of the IVE systems and maintenance of the data within the business.

The management of change may be simplified by dividing the process into a logical sequence of steps arranged on either a functional basis (such as human resources) or on an operational basis (such as design process models). Explicit milestones can help to monitor and manage progression.

Providers

- By allocating responsibility for managing its business interfaces with other value system members to an individual within the organisation (a project or design manager, for example), a provider can ensure that the implementation of its IVE strategy remains aligned with that of other value system members.

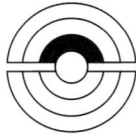

How well are we doing?

This stage is concerned with monitoring the effect your strategy is having, so that you can take corrective action, if required.

Determine Feedback Required
Your strategy should be adaptable during its deployment to allow for any obstacles encountered. Identify the key measures required, especially those associated with the strategic targets.

Establish Feedback Mechanisms
The maturity of IVE business domain competencies should be kept under review. The outcomes should be monitored and the strategy revised if it is not having the intended effect and/or to take account of any changes that may have occurred in the business environment.

Gather and Respond to Feedback
Once IVE is established and underway, it is important that it remains appropriate to the operating environment. In a similar manner to responding to change when establishing IVE practice, this continuous review helps the organisation determine if its emerging IVE practice remains compatible with the latest operating needs of the business.

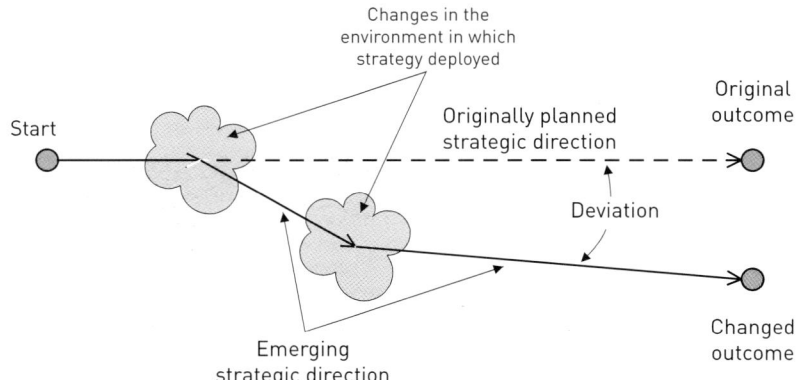

Strategic Business Development

Providers

- Continuously reviewing IVE practice gives providers the means to ensure their practice remains appropriate to the needs of the receivers they support. If a provider's operating environment is particularly volatile, continuous review will increase a provider's ability to maintain its presence in a value system.

Receivers

- By monitoring the environment in which the organisation or value system functions, a receiver can tell when it is necessary to adjust the flow of design information between providers and receivers to redistribute design tasks between them. Modelling Business Design Processes (BO1) can be used to make these changes.

BT1

Applying Process Management in the Business

Provider benefits

- By adopting a process view of their business and project operations, providers will be able to define their high-level and detailed level processes, creating the opportunity to develop a common language that can be used by the whole organisation, and between other supply network members.

Purpose

By adopting a process view of their business and project design operations, organisations will be able to align their businesses in ways that will improve their overall design process management. This practice introduces the tactical response to the needs of the business, as outlined in the business domain strategy, Planning Design Process Management (BS1).

Summary

Having identified a need to adopt a process view, ICD organisations will be required to determine the level of their understanding of design processes. This will enable the organisation to plan what needs to be done in order to derive, or further develop, a representative model of their business and subsequent project design processes. ICD organisations should use a 'high-level' process framework to manage their business and project operations, providing them with the context in which to drill down and develop a detailed understanding of their design processes. The types of process models developed and the level of detail captured should be reviewed so that the application of these processes can be captured and fed back to the organisation. The business strategy should be reviewed leading to the streamlining of business and project domain processes based upon known competencies and a rational approach to risk allocation.

By establishing a coherent internal view of design processes, organisations should be able to review the implications of externally defined processes more easily.

Receiver benefits

- Receivers can consolidate their understandings of their organisational business and project processes, thus streamlining internal processes and improving alignment.

- Receivers will be able to communicate their business and project needs to other supply network members, thereby improving their external alignment.

Outline Procedure

1. Determine Results Required
2. Plan and Develop Approaches
3. Deploy Approaches
4. Assess and Review Approaches

Requirements and Resources

- Organisations should have adopted the ICD principle of Adopting Process Management and the practice Planning Design Process Management (BS1).
- Organisations must ensure that suitable individuals are engaged to champion and manage the process management activity. This will include key decision-making regarding the extent of modelling required.
- Informed individuals are required to determine the existing understanding of design processes.

Related Practices

- ✔ **BS1** Planning Design Process Management
- ☐ **BS2** Planning Supply Chain Management Business Practice
- ☐ **BS3** Planning the Implementation of Integral Value Engineering across the Business

- ☐ **BT1** Applying Process Management in the Business
- ☐ **BT2** Aligning Supply Networks
- ☐ **BT3** Applying Supply Chain Management in the Business
- ☐ **BT4** Applying Integral Value Engineering in the Business
- ☐ **BT5** Conducting a Value Survey
- ☐ **BT6** Performing an ADePT Review

- ✔ **BO1** Modelling Business Design Processes
- ✔ **BO2** Auditing the Supply Network
- ✔ **BO3** Auditing the Supply Network for Technical Competence
- ☐ **BO4** Gathering Value-adding Feedback from Projects

- ☐ **PS1** Planning Project Design Management
- ☐ **PS2** Planning Supply Chain Management Project Practice
- ☐ **PS3** Planning Integral Value Engineering Project Practice

- ☐ **PT1** Applying Design Management Practices
- ☐ **PT2** Applying Supply Chain Management to a Project
- ☐ **PT3** Selecting Supply Chain Members at the Project Level
- ☐ **PT4** Implementing Integral Value Engineering on a Project

- ☐ **PO1** Applying ADePT to Design Management
- ☐ **PO2** Applying DePlan to Design Management
- ☐ **PO3** Modelling Project Design Processes
- ☐ **PO4** Assembling Supply Chains
- ☐ **PO5** Applying Value-adding Tools to Design Problems

Determine Results Required

Each ICD organisation must communicate to employees the need to understand and model design processes. Collaborating parties need to recognise that design is an information intensive process, where transparent communication among designers and supply network members is critical to the successful delivery of a design solution.

The tactical objectives of defining process management should be to establish a common framework and language to secure alignment of processes both internally and externally.

Internal process management should provide a framework for the business to operate a consistent approach to design information exchange throughout its operations.

External alignment of processes can be more problematic since disparate organisations can have widely differing views on their role within a process. In establishing a supply network for serial projects it may be possible to get receivers and providers strategically aligned behind a common view of the processes. For specific project design chains or wider supplier networks, it may be simpler to establish an alignment of stage-gates only where collaborators can ensure that design information, cost planning and risk analysis are aligned at key points within the process.

The approach to process management should be addressed through training, knowledge capture and feedback. Collaborating organisations should be aware of the difference between the high-level process frameworks (that represent their generic business and project processes) and the detailed level design process models (that represent the design activity and associated information flows), so that they can respond to the business strategy in the most appropriate manner. An understanding of information requirements should enable the transition from traditional 'push' distribution of documents to 'pull' based information exchange.

Providers

- By establishing a coherent view of their internal design processes, providers can ensure that they respond to multiple receivers in a consistent manner. Establishment of a process understanding will help to ensure alignment of expectations at key stage-gates with receivers.

Receivers

- Receivers can use the definition of internal processes to understand their information needs and to align these with key providers. External processes may be strategically aligned for serial projects with a supply network by establishing a single coherent design process. Where a single design process cannot be established then an alignment of design information, cost planning and risk assessment at key stage-gates is essential.

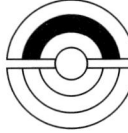

Plan and Develop Approaches

Having established the need to model design processes, each ICD organisation must determine the current understanding of their internal design process. This is to ensure that current understandings are consolidated and, in some cases, further developed to encompass the high-level process framework. The detailed level design processes that represent their business and project operations should also be reviewed.

An organisation may audit their own process management competency in terms of SCM and technical design capability by applying the two business domain operations: Auditing the Supply Network (BO2) and Auditing the Supply Network for Technical Competence (BO3). If an organisation has little or no current design process definition, then it should implement the business operation of Modelling Business Design Processes (BO1), which will provide them with the necessary tools to explore and derive a process framework and detailed level design process models.

Providers

- Providers may have to respond to multiple process definitions from multiple receivers with the risk of confusion in internal processes. A coherent internal process will help support a review of each of the receiver's needs enabling commonalities and differences to be examined and recognised.

Receivers

- Sharing the development and definition of design processes with key providers will help to establish a common view of design process, reducing gaps and overlaps between the collaborating parties.

Deploy Approaches

Once the level of understanding of existing processes has been determined, ICD organisations must select an appropriate tool to model their business and project design processes. When considering the high-level process framework that represents the overall business and project process, each ICD organisation may choose from a number of existing generic frameworks available to them (for example, the Project Process Protocol, see, www.processprotocol.com). These generic frameworks may be customised to represent the operations of the organisation, enabling them to build a representation that accurately reflects their operational processes. If an organisation has already developed their high-level process framework, then they will need to select an appropriate tool to assist them to drill down and further define their design processes. The development of a generic process framework and design process model can be achieved by adopting the practice Modelling Business Design Processes (BO1).

Having identified a suitable process framework, each ICD organisation will need to apply the framework to their business operations, where it can be modified to be representative of their business and project processes (see, Modelling Business Design Processes (BO1)). The development of process modelling within a business may require a consistent approach to process definition, which can be considerably enhanced by the use of an appropriate software tool to capture and represent these processes. Training in process modelling, including coaching, pilot programmes and feedback, will help encourage uptake. ICD organisations should be able to devise their own version of a process framework, enabling them to explore their processes in greater detail.

An externally focussed process framework can be developed with potential collaborators by seeking alignment of stage-gates and information flows. A fully aligned supply network will be able to rationalise its processes and recognise particular trading conditions between its members.

Process maps and models can be communicated in a variety of ways. Consider using graphical images to present the structure and nature of process maps in a clear and concise manner. High-level maps can be defined graphically and conveyed through posters, paper documents and company intranets. Specific modelling tools can be used to define process maps, including specialist software to assist in their development. With increasing complexity and connectivity in modelling, detailed level process maps can be enhanced by using modelling software to ensure integrity of data flow and hierarchies within processes. When developing models it is important for modellers to employ a set of templates to ensure that a standard style and process representation is produced.

Assess and Review Approaches

Methods of monitoring the use of process management resources within the organisation need to be established. This will ensure that the monitoring required to check how these resources are being used and that this is in line with the organisation's strategic objectives.

There are two elements to assessing and reviewing the effectiveness of approaches to improve process management:

1. Feedback mechanisms need to be in place to capture any lessons learned (see Modelling Business Design Processes (BO1)).

2. The monitoring exercise also needs to determine when processes become redundant (due to a change in the types of projects it undertakes, or the technical design problems arising within them) and pass this information back to the business domain.

Particular project circumstances should be addressed to determine whether the generic design processes can be applied. This would include a review of the appropriateness of the generic project process when working internationally or when responding to particular client requirements. It is particularly important to establish a coherent view of design processes when working in a multicultural environment.

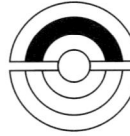

BT2

Aligning Supply Networks

Provider benefits

- Providers can improve their ability to work with, and learn from, other organisations through supply network alignment and this can happen before any projects work begins.

- Following on from the above, alignment also means faster earning curves and, thus, faster design work.

- The investment in aligning companies has the effect of binding providers to receivers and, thus, acts to increase security for the provider.

- Alignment increases the understanding the organisations have of each other. This makes behaviour more predictable and, thus, reduces the risks of things going wrong.

Receiver benefits

- As with providers, receivers are also bound by alignment to their providers, and need to work within the network to realise the investment they have been making in bringing about the network.

- Again, as with providers, the alignment process shortens the learning curve and means faster design.

- The establishment of minimum SCM maturity levels ensures that known levels of competence exist in the design chain.

Purpose

This practice aligns the business and the project processes of supply network members to allow provider organisations to work together as part of a design chain. Aligning the processes means creating common cultures and this greatly enhances the effectiveness of collaborative working.

Summary

The practice of aligning supply networks needs to be understood in conjunction with two other practices: Auditing the Supply Network (B02) and Modelling Business Design Processes (B01). The former details how existing supply networks can be improved and the latter describes how design efficiencies can be created by identifying generic processes - something it is particularly important to co-ordinate across the supply network.

There are a variety of ways of aligning a receiver with its supporting providers. Some of the more common ways are as follows:

- Using a SCM Maturity Matrix (how to do this is outlined in Auditing the Supply Network (B02)). This details organisations by the skills they possess and how competently they can exercise them. It allows receivers establishing supply networks to improve the performance of particular areas by means of an SCM audit.

- Minimum SCM maturity levels can be used as the entry qualification for the supply network membership. This will help a supply network maintain a common level of maturity among its members when new members are introduced.

Outline Procedure

The alignment of supply networks involves a number of stages. The SCM Maturity Matrix is the first step but, with this in place, the order in which the next stages are completed does not matter.

1. Compiling an SCM maturity matrix
2. Using the SCM audit to check alignment
3. Using design process models to align organisations
4. Aligning key SCM practices
5. Specifying minimum requirements for supply network membership
6. Formalising project practices

Requirements and Resources

To profit from supply network alignment, a company needs to have given consideration to SCM in planning its business objectives (see Modelling Business Design Processes (BO1) and Auditing the Supply Network (BO2)).

Related Practices

☐ **BS1** Planning Design Process Management
✓ **BS2** Planning Supply Chain Management Business Practice
☐ **BS3** Planning the Implementation of Integral Value Engineering across the Business

☐ **BT1** Applying Process Management in the Business
☐ **BT2** Aligning Supply Networks
☐ **BT3** Applying Supply Chain Management in the Business
☐ **BT4** Applying Integral Value Engineering in the Business
☐ **BT5** Conducting a Value Survey
☐ **BT6** Performing an ADePT Review

✓ **BO1** Modelling Business Design Processes
✓ **BO2** Auditing the Supply Network
☐ **BO3** Auditing the Supply Network for Technical Competence
☐ **BO4** Gathering Value-adding Feedback from Projects

☐ **PS1** Planning Project Design Management
☐ **PS2** Planning Supply Chain Management Project Practice
☐ **PS3** Planning Integral Value Engineering Project Practice

☐ **PT1** Applying Design Management Practices
☐ **PT2** Applying Supply Chain Management to a Project
☐ **PT3** Selecting Supply Chain Members at the Project Level
☐ **PT4** Implementing Integral Value Engineering on a Project

☐ **PO1** Applying ADePT to Design Management
☐ **PO2** Applying DePlan to Design Management
☐ **PO3** Modelling Project Design Processes
☐ **PO4** Assembling Supply Chains
☐ **PO5** Applying Value-adding Tools to Design Problems

Compiling an SCM Maturity Matrix

As we have seen, the SCM Maturity Matrix forms the basis from which supply network members can be audited. Determining which practices need to be included in the matrix is an important part of the process and is described in some detail in Auditing the Supply Network (BO2). However, it is worth noting here that practices can be selected from a variety of sources, including project reviews and business reviews - which might be conducted by Planning Supply Chain Management Business Practice (BS2). Where a supply network already exists and a matrix is being revised or recompiled, practices can be selected by means of the audit process.

Putting together an SCM Maturity Matrix involves a number of stages, which are best conducted in a workshop. This creates opportunities to benefit from open discussion and facilitation (by an individual who is appropriately skilled and who may be a member of an organisation in the supply network or an independent third party consultant):

- A template of the matrix needs to be completed and checks made that the practices that have been entered are all that the organisation's SCM strategy requires. Where the audit is being reapplied (as opposed to its initial implementation), the maturity levels at which each practice is entered in the matrix will need to be verified and amended as required to take account of any improvements in overall supply network maturity since the last audit.

- Review the content of the generic matrix. Add additional practices relevant to the types of work your supply network currently undertakes (or which it plans to undertake as it seeks to reposition its business in the market).

- Delete those practices that do not contribute towards the receiver's SCM strategy.

- Allocate SCM practices to appropriate levels of maturity within the matrix.

Providers

- Supply networks begin by reflecting the needs of the receiver. As they develop and the needs of the receiver become aligned with those of the providers, the supply network becomes increasing integrated and, thus, more useful to providers.

Receivers

- When initially assembling a supply network, the SCM Maturity Matrix, as discussed above, must be generated using the practice template. When used subsequently, and when audits have taken place, the matrix can be revised by the receiver to allow it to track the direction the business is taking, and identify any new SCM practices that the network needs to acquire or that may have been identified when completing the audit.

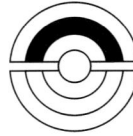

Using the SCM Audit to Check Alignment

Supply network audits are conducted to periodically check how well providers and receivers are aligned. Comparing the SCM maturity profiles of each (see figure) is a way of identifying whether any individual member is significantly different from the rest of the network and also which organisations are the most similar.

The data generated by the SCM audit can be used to set minimum or key requirements for the network, as well as detailing where practices are widely used. Knowing which practices are widely used then becomes the basis for developing generic processes.

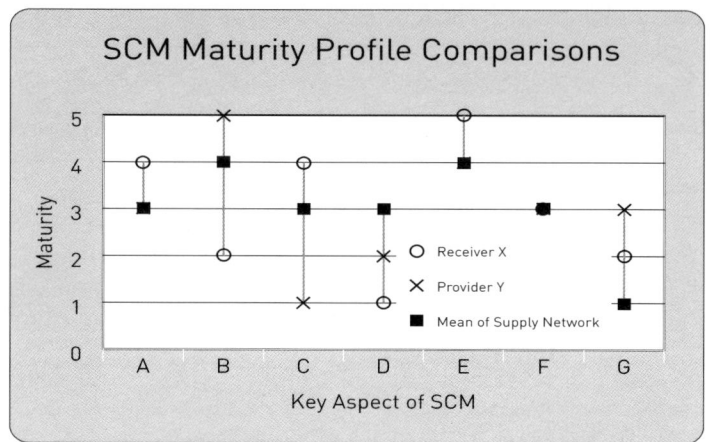

Providers

- Over time, providers (independently of receivers) can use the supply network to standardise and develop generic processes and practices.

Using Design Process Models to Align Organisations

Agreeing the distribution of design processes between supply network members allows more detailed protocols (for example, computer-aided design (CAD) protocols) to be defined and, where the processes are generic, this can happen independently of any project. As we have seen, one of the ways generic processes can be identified is by modelling the whole design process of each project. See Modelling Business Design Processes (BO1).

Aligning Key SCM Practices

The process of identifying which key SCM practices to select in order to develop the maturity of the supply network will lead to the SCM competencies of both receivers and providers converging. Through this convergence process, providers are in a position to influence the development of SCM practice among all network members - for example, one provider might recognise the need for greater IT integration, and by developing expertise in this area, influences and informs other members of the network.

Specifying Minimum Requirements for Supply Network Membership

As well as identifying key areas for development, the SCM audit can, as we have seen, be used by a receiver to set minimum SCM maturity criteria for supply network membership. In the early stages of establishing a supply network, a provider's SCM maturity profile may provide a sufficient measurement of its ability for it to be included in a network. However, as the network develops, it may be necessary for a receiver to stipulate more specific practice maturity levels and, in this way, steadily improve the skill base of the supply network.

Receivers

- Setting minimum performance standards can cause problems. Performance standards are raised with the threat of ejection form the network (and possible corresponding loss of work) and this needs to be real to be effective. However, when this begins to undermine the trust in the relationship between receivers and providers. Also, where providers have been ejected, there are times, because of their specialist skills (even if not of the desired standard), when they still need to be used. This is another factor that can undermine provider confidence in the network.

Formalising Project Practices

As we have seen, formalising project practices is one of the ways by which providers and receivers can be aligned. When networks begin, many project practices develop in an unplanned fashion. With time, and through supply network forums, for these practices become formalised. This allows generic project procedures and guidelines to be established and incorporated into other ICD practices, such as:

- planning supply chain project practice;
- assembling project supply chains;
- applying supply chain project practice; and
- assembling supply chains;

BT3

Applying Supply Chain Management in the Business

Provider benefits

- The formalisation of practices, which occur with SCM, makes the relationship between provider and receiver more predictable.
- Linked to the above, SCM improves communication within the network.
- Closer working relationships result from the development of dedicated teams to serve particular providers.
- The use of strategic purchasing by a receiver creates and stabilises the business environment enabling partnering with providers, which supports their design activities.

Receiver benefits

- Strategic purchasing encourages receivers to exert greater influence over providers by building closer relationships and making both increasingly dependent on each other.
- Selection of appropriate implementation strategies allows organisations to modify how they approach new working practices.
- Improves communication with, and between, organisations.
- The formalisation of practices, as noted in the provider benefits, makes for better working relationships and allows innovation, when it occurs, to be communicated to other team members more reliably.

Purpose

Supply Chain Management (SCM) helps companies establish working practices within their ICD approach, which structure their use of supply networks to inform the assembly of project supply chains and, therefore, design chains.

Summary

This practice, which covers a number of issues that companies instigating SCM should consider, needs to be read in conjunction with Planning Supply Chain Management Business Practice (BS2)). Particular issues include:

- reviewing your SCM strategy;
- the role, application and mechanisms of strategic purchasing;
- selecting appropriate implementation strategy(ies) for use in both the business domain and the project domain;
- formalising business practices - making explicit and formal practices which may already be widespread;
- business partnering - building long-term formal and informal relationships within the business domain;
- how the organisation communicates, both internally and with other organisations, within the project and also outside of them; and
- how the organisations define generic design work packages for retention in the business domain and application as strategic devices in future projects.

While many of these issues are not unique to SCM, some (such as business partnering and issues of communication, for example) are especially important when it comes to considering how supply networks are to be established and maintained. Receivers (and indeed providers who may act as receivers within their own supply networks) need to pay particular attention to these issues in order to obtain the maximum benefit of adopting SCM practices.

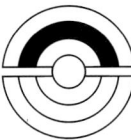

Outline Procedure

There is no one way to apply SCM to a business, as a variety of approaches can be effective. Seven approaches are described within the following procedure:

1. Determine results required
2. Plan and develop approaches
3. Deploy approach - strategic purchasing
4. Deploy approach - formalising practices
5. Deploy approach - business partnering
6. Deploy approach - improving internal communication
7. Deploy approach - improving external communication
8. Deploy approach - managing knowledge
9. Deploy approach - using dedicated project staff
10. Assess and review approaches

Requirements and Resources

- Before applying SCM, an organisation needs to have done a certain amount of planning. Details of what needs to be done are given in Planning Supply Chain Management Business Practice (BS2).
- The resources required for SCM are not special but depend on the environment in which it is being introduced.
- Better use of generic work packages is one of the advantages of SCM. However, to take full advantage, an organisation needs to have first modelled its design process (see Modelling Business Design Processes (BO1)).

Related Practices

- ☐ **BS1** Planning Design Process Management
- ☑ **BS2** Planning Supply Chain Management Business Practice
- ☐ **BS3** Planning the Implementation of Integral Value Engineering across the Business

- ☐ **BT1** Applying Process Management in the Business
- ☐ **BT2** Aligning Supply Networks
- **BT3** Applying Supply Chain Management in the Business
- ☐ **BT4** Applying Integral Value Engineering in the Business
- ☐ **BT5** Conducting a Value Survey
- ☐ **BT6** Performing an ADePT Review

- ☑ **BO1** Modelling Business Design Processes
- ☑ **BO2** Auditing the Supply Network
- ☑ **BO3** Auditing the Supply Network for Technical Competence
- ☐ **BO4** Gathering Value-adding Feedback from Projects

- ☐ **PS1** Planning Project Design Management
- ☐ **PS2** Planning Supply Chain Management Project Practice
- ☐ **PS3** Planning Integral Value Engineering Project Practice

- ☐ **PT1** Applying Design Management Practices
- ☐ **PT2** Applying Supply Chain Management to a Project
- ☐ **PT3** Selecting Supply Chain Members at the Project Level
- ☐ **PT4** Implementing Integral Value Engineering on a Project

- ☐ **PO1** Applying ADePT to Design Management
- ☐ **PO2** Applying DePlan to Design Management
- ☐ **PO3** Modelling Project Design Processes
- ☑ **PO4** Assembling Supply Chains
- ☐ **PO5** Applying Value-adding Tools to Design Problems

Determine Results Required

An organisation must regularly review its SCM business strategy to ensure it remains able to support the organisation's wider business needs, while also allowing the company to function effectively in its current environment. Such a review should:

- examine the current position of the business with regard to the supply network and the intended future SCM direction, checking, for instance, to see whether the existing network capabilities reflect client demands;

- look at those elements of the plan concerned with the development of SCM practice within the organisation - the organisation needs to be clear that these are adequate for its wider business needs, if not, the organisation will need to revise its SCM strategy to bring it back on track; and

- compare the SCM practices the company possess with those the supply network requires (this includes not just skills but particular competencies in those skills) to determine whether shortfalls exist and, if so, how they can be corrected.

The organisation's SCM strategy should be reviewed after each supply network audit (see Auditing the Supply Network (BO2)). The frequency of review should be increased if significant changes are anticipated in the organisation's operating environment, or if that environment becomes unstable.

Plan and Develop Approaches

SCM implementation must include appropriate education and training. Part of this education is a general awareness and understanding, across the organisation, of what SCM entails. There are two main ways to do this:

1. A Directive Approach. Top-down implementation from senior management to project staff may indicate when and how new practices and tools are to be used. In such an approach, new rules or procedures are established and staff trained accordingly.

2. An Adaptive Approach. Tools and practices are made available within the organisation, together with outline guidance. Project personnel are expected to experiment with these tools. Management then supports any innovation that emerges and, where seen to be beneficial, encourages adoption of new methods. This approach, because it has been nurtured from the shop-floor, is more likely to get employee buy-in than one that imposes new methods from above.

However, SCM implementation strategies must balance both these approaches to accommodate differing project and business conditions. The mixture of adaptive and directive implementation is determined by the existing organisation culture. Where organisations are more design-led, the adaptive approach may be the more useful (but may still encounter resistance from designers).

Deploy Approach - Strategic Purchasing

Strategic purchasing uses a receiver's long-term relationship with its key providers to control the business environment in which projects are delivered. A receiver should use strategic purchasing to develop an ongoing procurement programme (and relationships with provider business partners).

Strategic purchasing is used to reduce variability and promote consistency across supply and design chains. It does this in the following ways:

- Reducing the number of available suppliers to individual projects.

- Using performance benchmarking to select organisations for the supply network. This may be achieved using project reviews (described in Assembling Supply Chains (PO4)); by other measures such as a SCM Maturity Index (described in Auditing the Supply Network (BO2) or by the Key Performance Indicators developed by the Construction Best Practice Programme[1] (see Auditing the Supply Network for Technical Competence (BO3)).

Continued on next page

[1] For more information see www.cbpp.org.uk

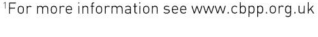

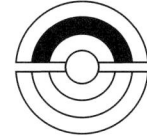

Continued from previous page

- Acting as a gatekeeper to the supply network members to maintain the status and benefits to providers of membership of a supply network. Strategic purchasing acts as a filter between available providers and the projects by creating barriers to entry into design chains (such as membership of the supply network) through imposing limits and conditions surrounding the appointment of providers to design chains. This does not exclude the use of other organisations who are not members of a supply network from working on individual projects, but requires that limits be set on the ability of project teams to operate autonomously and appoint organisations without consideration of wider business interests.
- Selecting or classifying providers in a particular way, such as on the basis of value and/or risk. High-risk members of the supply chain are also typically members of the design chain, as these are the organisations where SCM can provide the greatest benefits in managing cost, quality and programme. This type of selection can be done by using the strategic purchasing function to negotiate annual price agreements.

Providers

- Strategic (rather than purely price-driven) procurement increases trust. Although strategic purchasing mechanisms may not guarantee workload they create frameworks within which the appointment of supply network members to projects takes place and allows providers to understand how they are likely to be affected by a receiver's procurement policy. The trust comes through the ability of the receiver to ensure the project personnel comply with the terms and the spirit of the business partnership, and that the providers will get a 'fair crack of the whip'.

Receivers

- Traditionally, receivers view purchasing as a response to design and not as a strategic instrument. However, as the level of a receiver's SCM maturity grows, the view often shifts and purchasing comes to be seen less about project needs and more as a way to influence strategic relationships.
- Annual price agreements, as we have seen, reduce price variability. Rebates also provide a guaranteed margin and tangible benefit from maintaining long-term business relationships with strategic partners; thus, they give additional justification for adopting SCM practices.

Deploy Approach - Formalising Practices

Ways of working need to be formalised to allow them to be clearly communicated throughout an organisation and to those external organisations it works with. It is often easier to use existing systems of communication, standardisation and control (such as Quality Assurance, for example) to spread formal practices throughout an organisation. However, there is a risk that this will only have a limited impact on people's behaviour because of the tendency to see formal processes and practices as an impediment to an individual's (as opposed to the organisation's) preferred way of working. Ad hoc processes can also be used. These might include using individuals as 'ambassadors' to take the message between parts of the business, or between projects, as they move around the organisation. This technique might be better at changing behaviour but may only be partially effective, as the ambassadors may not come into contact with everybody, or with many people particularly quickly.

A practice is formalised in three stages - identification, communication and implementation:
1. The practice needs to be identified, noted and recorded.
2. The identified practice needs to be communicated through guidance notes or procedures manuals.
3. The documented practice then needs to be implemented, which, in turn, needs to be tested - to see whether it complies with an existing QA system for example.

Deploy Approach - Business Partnering

By partnering[2] individual providers and receivers can develop a mutual understanding that will develop over time and span many projects. Such relationships, largely built on trust, have many advantages; most importantly they avoid the need for business relationships to be re-established with each new project[3] and they allow shared working principles, practices and methods to be established and carried over from one job to the next.

Within a partnering arrangement, organisations align their working methods with one another. This gives them an effective way to combine their expertise, making the sum greater than the parts. In addition, the simplification of the working relationship creates economies of scale (e.g. through the reduction of transaction costs) and the savings that result can either be retained as profit, or used as a means of securing (more) work by making the fees more competitive. Partnering uses both formal and informal arrangements. Often the relationships between organisations, which evolve over time, can work in the same way as a formal partnering agreement. On other occasions, however, companies may wish to define their relationships formally from the outset.

Receivers

- When establishing partnering relationships, a balance must be struck between the benefits of having a long-term relationship and the need to prevent such relationships from stagnating. A common way firms manage this balance is to change business partners every five years or so to maintain commitment, innovation and vitality.

- Regular reviews must be implemented so that issues, such as workload (future and current), performance and communication, can be discussed outside of the project environment.

Deploy Approach - Improving Internal Communication

Communication is vital for SCM, particularly when strategic change - often viewed as a threat by the workforce - is being contemplated. With poor communication, organisations can risk losing valued staff by failing to get the right message across about the changes being planned. This can then have a knock-on effect and undermine the organisation's ability to introduce change.

SCM practice needs to be communicated to the workforce in a consistent manner. This can be done in a number of ways:

- Individuals, who are already familiar with SCM, can be used as champions. They can be given influential roles, such as on steering groups or in management committees, and used to co-ordinate the way SCM is introduced.

- New staff for specific SCM roles, such as supply chain manager, can be taken on. This reinforces the message informally that the company is committed to the concept.

- Senior and middle managers can include regular SCM updates within their normal briefing meetings with the workforce and use the opportunity to introduce any changes needed.

Other channels of communication - intranet, presentations, internal newsletters and informal sessions - should be used, where possible, to give a consistent message.

[2]Bennett, J. and Jeyes, S. (1998) *The Seven Pillars of Partnering*, Thomas Telford, London

[3]Balow, J., Cohen, M., Jashapara, A., and Simpson, Y. (1997) *Towards positive partnering; Revealing the realities in the construction industry*, The Policy Press, Bistol

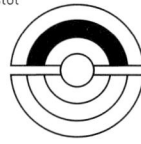

Deploy Approach - Improving External Communication

Communications with providers in the supply network are just as important. As has previously been noted, one of the benefits of the supply network is the opportunity it creates for providers and receivers to communicate with one another outside of any project environment. Communicating with the supply network can be done relatively easily, with intranets (which give limited access to each member's business information) with 'supplier suppers' (where members meet semi-socially outside of projects) and with newsletters and in-house magazines.

Keeping one another up-to-date is important if the supply network is to flourish. The sort of information that needs to be exchanged regularly includes such things as the details of an organisation's existing activities and the future direction in which it intends to go. In more advanced supply networks, this sort of exchange can also be a source of future opportunities for network members. Together, companies may be able to see possibilities that are within the scope of the network but which are too big for any one member.

Providers

- Providers can take the initiative and use the range of communication channels already discussed to build relationships between organisations. Such channels include client evenings, intranets and newsletters. All go to create a sense of openness, which is necessary for effective collaboration.

Receivers

- Receivers can extend existing means of internal communication - newsletters and intranets - to the external provider network, making the members of the network feel more trusted and closer to the receiver.
- The receiver, who needs to build a small number of close relationships, can use workshops and 'supplier suppers' as a way of doing this. These, however, need to be targeted affairs and cannot be too big or they lose their effect.

Deploy Approach - Managing Knowledge

SCM helps organisations overcome the fragmentation of knowledge that exists in construction when design chains are assembled from supply chains. There are two strategies that can be used:

1. Use 'design gurus' with specialist technical knowledge. These individuals can be slotted into projects as required to provide support at key points. This avoids the need for organisations to field experts in every area and gives all organisations in the network access to specialist knowledge.

2. Develop generic work packages (see Modelling Business Design Processes (B01)). These templates or typified work packages, which might comprise ductwork or profiled metal cladding, reflect systems or components that are common across projects and can act as a starting point for determining work packages for individual projects.

Deploy Approach - Using Dedicated Project Staff

Where resources allow, key personnel can be identified who may repeatedly work with other organisations. This allows expertise and familiarity with the procedures and practices to be built up. This consistency reduces the risk of misunderstandings arising between receivers and providers, while also acting as a pool of experience to which other people within the organisation can refer. This may result in a variety of arrangements:

1. Key relationship managers, where a relationship is overviewed by an individual.
2. Core teams; groups of personnel who can serve or work with a specific client, receiver or provider.
3. Individual designers who may have more experience of working with another organisation and so would be best placed to work with them on future projects.

Providers

- Such an approach is dependent on the provider's resources. Individuals or core teams may need to deal with several different client types, while overlapping with other core teams or individuals. Where this occurs, providers will need to ensure that resources can still be allocated to meet the current workload.

Receivers

- Receivers need to keep in mind that key personnel may not want to be 'pigeon-holed' into a particular type of work. Designers, in particular, may require variety in their workload in the types of building they work on and the systems or components they design while others may prefer to specialise.
- Receivers must ensure that, as knowledge flows into the organisation from the network, their own staff get the experience they need. There is a danger the network can supply knowledge more quickly than a workforce can assimilate it. Where this happens, staff can become frustrated and ineffective.

Assess and Review Approaches

The approaches used need to be assessed and reviewed to ensure they are achieving the objectives of the SCM strategy. In part, this should be carried out in conjunction with any review of the SCM strategy and after any audit of the network. This provides the opportunity to review and assess the SCM practices that have been developed to be appraised in the light of the approaches used.

BT4

Applying Integral Value Engineering in the Business

Provider benefits

- Providers benefit from IVE because it is a means for establishing a common basis for collaboration within the value system and of sharing resources and knowledge.

- By providing feedback, providers can ensure that they are not hampered by the way IVE resources are deployed.

Purpose

Integral Value Engineering (IVE) needs to be managed in the business domain to ensure that it corresponds with business objectives and is practised appropriately in projects. This involves tracing the benefit of a value-adding toolbox (and other IVE resources) - how they have been developed, how they have been aligned with a company's business objectives and how they have then been reapplied to projects.

Summary

This practice is concerned with making sure that the value-adding resources are made available so that an organisation can tackle projects in the way it wants to (defined in business objectives) and complete them effectively (defined as project objectives). It is also concerned with collecting feedback from new techniques being pioneered in projects. Compiling this information in the business domain lets subsequent projects reuse the IVE resources in new or improved ways. The resources can be varied in their nature and include any mechanism used to support IVE, including: a value-adding toolbox; a portfolio of value-adding tools; programmes of employee education and training; and methods of communicating project values to all project members.

Receiver benefits

- Receivers benefit from sharing IVE resources with their supporting providers within a managed value system, as it is a way of ensuring that everyone is conscious of how the techniques which they are collectively developing are helping them all achieve both their project and business objectives.

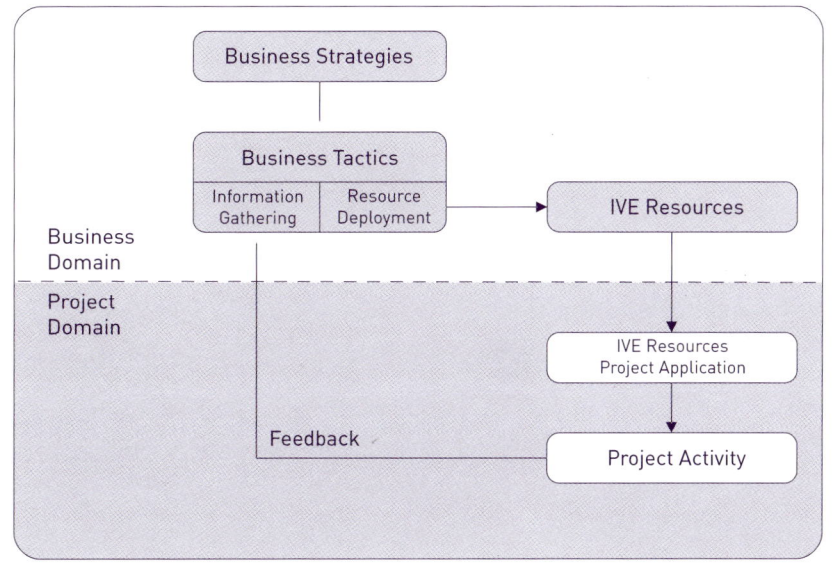

Outline Procedure

1. Determine Results Required
2. Plan and Develop Approaches
3. Deploy Approaches
4. Assess and Review Approaches

Requirements and Resources

- The various IVE resources deployed by this practice within the business (or value system) must be available within the organisation. This includes the prior decision to use and deploy these resources within the organisation.
- When IVE resources are deployed in the electronic form, appropriate IT provisions must have been made to create the communications infrastructure over which IVE resources are deployed. IVE resources may also contain appropriate technical, personnel and training provisions.

Related Practices

- ☐ **BS1** Planning Design Process Management
- ☐ **BS2** Planning Supply Chain Management Business Practice
- ☑ **BS3** Planning the Implementation of Integral Value Engineering across the Business

- ☐ **BT1** Applying Process Management in the Business
- ☐ **BT2** Aligning Supply Networks
- ☐ **BT3** Applying Supply Chain Management in the Business
- ☐ **BT4** Applying Integral Value Engineering in the Business
- ☐ **BT5** Conducting a Value Survey
- ☐ **BT6** Performing an ADePT Review

- ☐ **BO1** Modelling Business Design Processes
- ☐ **BO2** Auditing the Supply Network
- ☐ **BO3** Auditing the Supply Network for Technical Competence
- ☑ **BO4** Gathering Value-adding Feedback from Projects

- ☐ **PS1** Planning Project Design Management
- ☐ **PS2** Planning Supply Chain Management Project Practice
- ☐ **PS3** Planning Integral Value Engineering Project Practice

- ☐ **PT1** Applying Design Management Practices
- ☐ **PT2** Applying Supply Chain Management to a Project
- ☐ **PT3** Selecting Supply Chain Members at the Project Level
- ☐ **PT4** Implementing Integral Value Engineering on a Project

- ☐ **PO1** Applying ADePT to Design Management
- ☐ **PO2** Applying DePlan to Design Management
- ☐ **PO3** Modelling Project Design Processes
- ☐ **PO4** Assembling Supply Chains
- ☐ **PO5** Applying Value-adding Tools to Design Problems

Determine Results Required

IVE resources include all provisions and working tools created by a single organisation or developed and shared within a value system. These may include any mechanism used to support the creation of project value within design activity. These resources are applied during projects, but the infrastructure is maintained in the business domain to provide continuity and learning from project to project. They are focused on the maintenance and provision of a value-adding toolbox to disseminate corporate or value system knowledge of value-adding tools. This tactic seeks to ensure that appropriate IVE resources are available within an organisation (or value system) so those business goals and strategic targets related to IVE can be met.

Business objectives (for the company and for the value system) must be reviewed to assess whether the right sort of tools are being developed and whether there are sufficient training resources in place to keep everyone up-to-date on the latest techniques.

Plan and Develop Approaches

Your scope to plan and development approaches for the deployment of IVE resources within your organisation will be determined by the business strategy to which your use of this tactic responds. Depending on how the strategic goals of your business have been defined, you may have some choice of IVE resources, or this decision may have already been made at the strategic level.

The deployment of IVE resources must be carefully sequenced and planned. For example, a value-adding toolbox can only be used after value-adding tools have been identified. However, these, in turn, require a period of evaluation to determine which are most useful to the organisation (and should, therefore, be included in the toolbox). Further, neither tools or the toolbox will be integrated into the everyday of organisation members if they have not first gained insight into their benefits through education and training. This can be achieved through a combination of workshops and self-help documentation, such as instructions and worked examples.

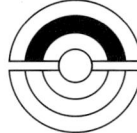

Deploy Approaches

The deployment of IVE resources must satisfy both business and project requirements, in accordance with the plan developed in the previous stage of this tactic. Sometimes the tactical long-term goals for the deployment and use of IVE resources within the business may conflict with the short-term goals and requirements of individual projects. An appropriate tactical response to establish an appropriate balance between the requirements of the two domains is required.

For example, the deployment of a value-adding toolbox as an IVE resource must balance business and project domain objectives, e.g.:

* in the business domain, the deployment of a value-adding toolbox would typically be concerned with ensuring that a useful portfolio of value-adding tools is maintained within the organisation and that it learns from their use; while

* in the project domain, the tactical provision of that IVE resource is solely concerned with making value-adding tools available for use.

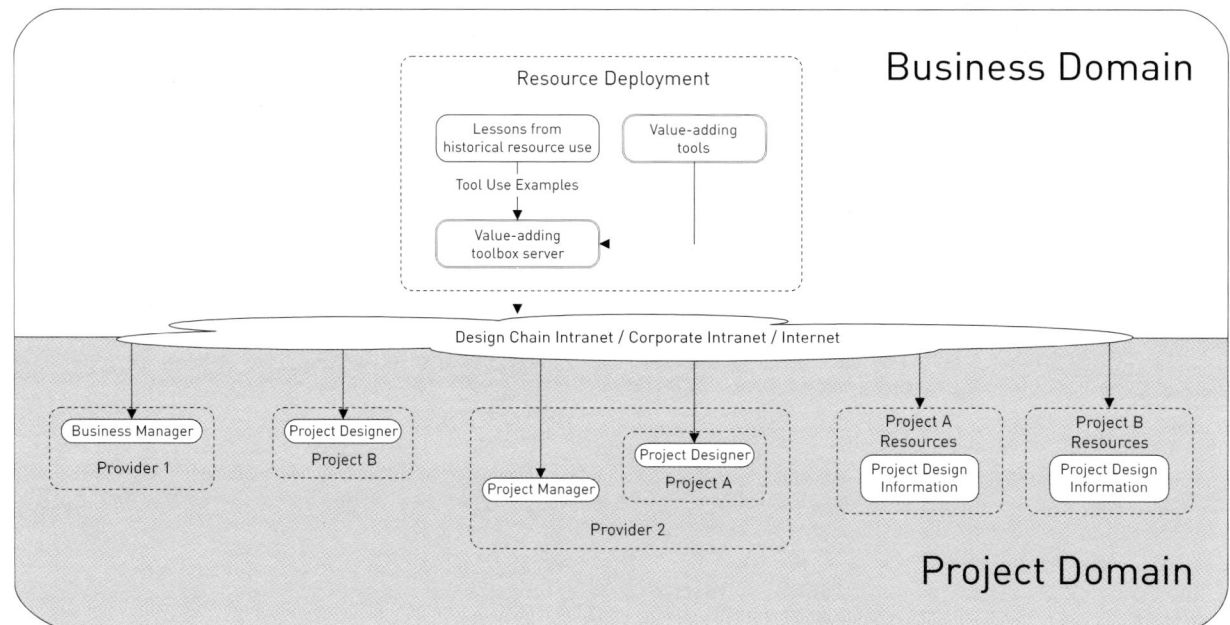

The most effective mechanisms will be through the internet and company intranets (or project extranets), which can deliver consistent information, facilitate concurrent collaborative working (e.g. using a net meeting) and ease collection of feedback.

Improvements in proficiency will require training in addition to experience. This applies to staff involved in the management of IVE and those applying the toolbox. A combination of workshop and distance learning (e.g. via an intranet) is likely to be most effective.

Providers

- Providers must make special provision to ensure that the tools they use can be easily integrated into the way they work, given that the resources to be used will typically be determined by a receiver elsewhere in the value system or project design chain.

Receivers

- Significant barriers can be apparent where IT and communications firewalls act as barriers to communication.

Assess and Review Approaches

Methods of monitoring the use of IVE resources within the organisation or value system need to be established. This will facilitate the monitoring required to check how these resources are being used and that this is in line with the organisation's (and value system's) strategic objectives.

There are two elements to assessing and reviewing the effectiveness of approaches to improve IVE practice:

1. Feedback mechanisms need to be in place to capture any lessons learned (see Gathering Value-adding Feedback from Projects (B04)).

2. The monitoring exercise also needs to determine when tools become redundant due to a change in the types of projects it undertakes or the technical design problems arising within them, for example, and pass this information back to the business domain, removing ineffective tools and adding new techniques.

BT5

Conducting a Value Survey

Provider benefits

- The survey creates an opportunity for receivers to review the unique skills and methods each provider contributes to projects and, thus, the role they play in any associated value system, including their ability to contribute to the provision of value to clients. Therefore, the survey provides an opportunity for individual providers to promote any relevant attributes they may have.

Purpose

A value survey gives a snapshot of the methods used by providers (typically within a value system) to address value within their work. This will determine the level of value-related competency and tools available within the system. The level of understanding of the nature of value provides the basis upon which these value-related ICD practices can be deployed.

Summary

A value survey gathers information about how providers understand the nature of value in construction projects, and the steps they take to contribute to its delivery. This is important as it helps the value system to which they belong develop a consistent approach to collaborative working and helps them understand the effectiveness of ICD practices, reinforcing their appeal to clients.

A receiver uses questionnaires and interviews to conduct the value survey. Although the survey is less structured than, for instance, Auditing the Supply Network (BO2), the issues it addresses are specific to each receiver (or value system) conducting the survey.

Receiver benefits

- Receivers are better able to plan business relationships, and the strategic directions they need to take, by first understanding the extent to which the understanding of value and value-delivery practices are consistent among their supporting providers.

- Surveys allow receivers to determine whether providers' understanding of value matches that necessary for project success.

- Surveys allow receivers to check that the methods providers use to deliver value in projects are compatible with their own.

Outline Procedure

A value survey is conducted by:
1. Administering the survey
2. Compiling questionnaires
3. Reviewing responses and identifying key issues
4. Conducting follow-up interviews

Requirements and Resources

- A value survey is most useful if the information generated by it is used in Planning the Integration of Integral Value Engineering across the Business (BS3).
- Receivers must assume responsibility for managing and administering each value survey. This is a business domain activity and, although the survey is administered at a low business level, strategic business managers must be available to act on the information it generates.
- Providers must use senior and/or informed individuals when they take part in surveys to ensure their organisation gives sufficiently detailed answers.

Related Practices

- ☐ **BS1** Planning Design Process Management
- ☐ **BS2** Planning Supply Chain Management Business Practice
- ☑ **BS3** Planning the Implementation of Integral Value Engineering across the Business

- ☐ **BT1** Applying Process Management in the Business
- ☐ **BT2** Aligning Supply Networks
- ☐ **BT3** Applying Supply Chain Management in the Business
- ☐ **BT4** Applying Integral Value Engineering in the Business
- ☐ **BT5** Conducting a Value Survey
- ☐ **BT6** Performing an ADePT Review

- ☐ **BO1** Modelling Business Design Processes
- ☑ **BO2** Auditing the Supply Network
- ☐ **BO3** Auditing the Supply Network for Technical Competence
- ☐ **BO4** Gathering Value-adding Feedback from Projects

- ☐ **PS1** Planning Project Design Management
- ☐ **PS2** Planning Supply Chain Management Project Practice
- ☑ **PS3** Planning Integral Value Engineering Project Practice

- ☐ **PT1** Applying Design Management Practices
- ☐ **PT2** Applying Supply Chain Management to a Project
- ☐ **PT3** Selecting Supply Chain Members at the Project Level
- ☐ **PT4** Implementing Integral Value Engineering on a Project

- ☐ **PO1** Applying ADePT to Design Management
- ☐ **PO2** Applying DePlan to Design Management
- ☐ **PO3** Modelling Project Design Processes
- ☐ **PO4** Assembling Supply Chains
- ☐ **PO5** Applying Value-adding Tools to Design Problems

Administering the Survey

The value survey comprises two elements: an initial questionnaire (which gathers information on current practices and tools) and a series of follow-up interviews (which obtain information on how practices are used in the understanding of the underlying issues).

The questionnaire needs to be simple, short and well structured. It is usually sent by post, but can be adapted for completion over the Internet. It investigates how providers understand the ways in which they contribute value through their involvement in projects. Each survey is customised to focus on the areas the receiver needs to examine in order to manage the delivery of value to clients from whole projects.

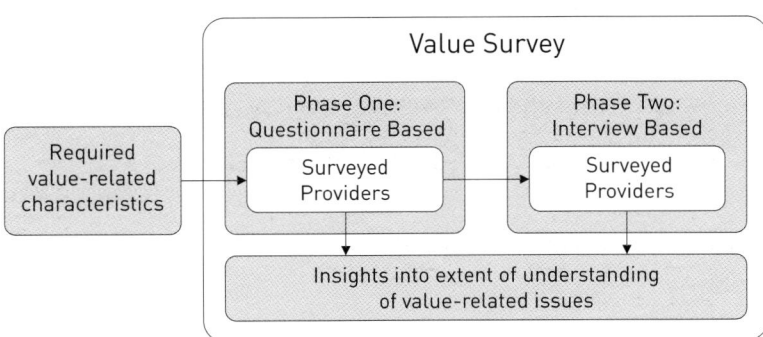

The questionnaire should address issues concerned with:
the current understanding of terminology used throughout the industry;

- the value-related practices (if any) used by the organisation;
- how they are used;
- the stage of projects to which they are applied; and
- the type of work to which they are applied.

The questionnaire should also contain unstructured, 'open' questions which provide respondents with opportunities to raise relevant issues with which they are concerned. The information gathered can be compared to determine the extent of consistency between organisations and, therefore, the extent of commonality of understanding of value-related issues.

Providers

- Survey responses must be based on historical practice and should not include anticipatory answers.

Receivers

- If the number of providers within a value system is large, a receiver can survey a smaller sample of value system members instead. Care must be taken to ensure that the sample reflects the composition of all providers within the system.
- Restricting the length of questionnaires to a single sheet of paper is a good way to keep them manageable. Questions should be straightforward and carefully structured, guiding the respondent from one to the next.

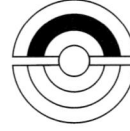

Compiling Questionnaires

To be effective, the questionnaires used in a value survey should be formatted to include three elements to address the following aspects of value-related issues:

1. An organisation's current understanding of value in general, including the elements of its practice and output that it considers to demonstrate the provision of 'good value' to others. This establishes the basic level of maturity of an organisation's understanding of the nature of value in its work and is largely determined by its experience.

2. Specific questions comparing an organisation's understanding of the meaning of value-related terms, methods and tools with their meaning that is generally accepted throughout the industry. This compares the organisation's understanding of specific elements of value-related issues with the benchmark established by the industry. If it is difficult to determine the industry understanding of a value issue, a value system may substitute its own definition as the benchmark against which the surveyed organisations are compared.

3. Exploratory questions that seek to identify any unique approaches to the delivery of value.

Reviewing Responses and Identifying Key Issues

An analysis of responses to the questionnaire will reveal how providers perceive and understand value. The level of consensus among providers within a value system can be determined, and, as noted earlier, this can be compared with the levels the receiver thinks necessary (as determined by Planning the Implementation of Integral Value Engineering across the Business (BS3) and its Planning Integral Value Engineering Project Practice (PS3)). In this way, the receiver will be able to check the stage of project progression, the type of process, and the type of information to which value-related practices are applied. These areas and any problems they throw up can then be investigated further in the follow-up interviews.

In the questionnaire, providers have the opportunity to give their views on how they see relationships with other companies in the value system, and whether they think some of these could be better managed. Although such views are given in an unstructured way, they nevertheless are an important source of information and are often investigated further at the interview stage.

Conducting Follow-up Interviews

Interviews are held with each provider to investigate each value-related issue further. The issues explored within each interview may either encourage or hinder value-adding contributions.

A semi-structured approach is often the best. This has the advantage of providing a framework that both guides the interviewer and helps in subsequent analysis and comparison of individual responses. It should also be sufficiently open-ended to allow the interviewee to raise issues of concern to them. Offering confidentiality will also reduce the concerns of those who wish to be critical or to admit to failures.

The results are summarised to give an insight into the informal issues that influence the way that providers work. If any issues are not fully understood, then further interviews should be conducted with those that raised the issue. If these determine the issue to be of sufficient importance, it may be necessary to amend and re-administer the questionnaire to find out the views of all surveyed organisations. If the issue is of lower importance, it may be ear-marked for investigation in a subsequent survey, implemented at a later stage in IVE deployment (see Planning the Implementation of Integral Value Engineering across the Business (BS3)).

Receivers

- Follow-up interviews are critical as they expose issues that may determine how the surveyed organisations should develop in the future. Although interviews may generate information in a format that can be difficult to analyse (notes, recordings, etc.), their findings must always be reviewed thoroughly to identify the critical issues.

BT6

Performing an ADePT Review

Provider benefits

- IVE can help resolve the design complexity represented by the groups of interdependent design from ADePT by creating appropriate collaborative problem-solving forums. These encourage organisations with specialised knowledge to introduce their expertise and suggest innovative solutions that would not otherwise come to the fore.

- The key features of the elements addressed by a block can be agreed by all parties prior to starting the design activity, thereby reducing abortive work.

Receiver benefits

- Receivers can use their knowledge of the key contribution of specialised providers to ensure that appropriate organisations are involved at an appropriate design stage.

Purpose

The Analytical Design Planning Technique (ADePT) is a method that supports the planning and management of design activity. ADePT clusters together design tasks that share iterative information flows, to form 'blocks' of interdependent tasks. The ADePT review examines the content of these blocks, in a representative sample of an organisation's projects, to determine whether they are generic (i.e. will re-occur in the majority of projects) and, if so, whether IVE can be used to tackle the technical issues that link them.

Summary

ADePT is a design management tool that groups interdependent design tasks in blocks. The blocks often contain design tasks from several design disciplines or organisations. IVE can be used to mange the complexity of iterative design and value-adding tools can create forums to help structure the collaboration of design chain members.

Groups (blocks) of interrelated design tasks that have arisen on past projects undertaken by a single organisation (or a single value system) over a period of time are analysed and the generic elements identified. The design content is then reviewed for IVE planning purposes and the blocks classified according to the design disciplines involved to increase the organisation's understanding.

To perform an ADePT review, an organisation must have first applied the ADePT to design management (see Applying ADePT to Design Management (PO1)). This practice provides the design task sequences that are examined by the ADePT review.

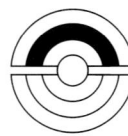

Outline Procedure

An ADePT review is performed by:
1. Identifying matrices for review
2. Identifying and describing design blocks
3. Compiling an IVE opportunities register
4. Disseminating to future projects

Requirements and Resources

- The ADePT review analyses information produced by the Analytical Design Planning Technique (ADePT). It therefore requires the practice to have been applied by the reviewing organisation to several projects. This in turn requires the practice of Applying ADePT to Design Management (PO1) to have been established before performing an ADePT review.
- The trained personnel and supporting software required by ADePT must also be available within the organisation to ensure that skills required to use ADePT, to generate the information analysed by this review, are available.

Related Practices

- ☐ **BS1** Planning Design Process Management
- ☐ **BS2** Planning Supply Chain Management Business Practice
- ☐ **BS3** Planning the Implementation of Integral Value Engineering across the Business

- ☐ **BT1** Applying Process Management in the Business
- ☑ **BT2** Aligning Supply Networks
- ☐ **BT3** Applying Supply Chain Management in the Business
- ☐ **BT4** Applying Integral Value Engineering in the Business
- ☐ **BT5** Conducting a Value Survey
- ☐ **BT6** Performing an ADePT Review

- ☑ **BO1** Modelling Business Design Processes
- ☐ **BO2** Auditing the Supply Network
- ☐ **BO3** Auditing the Supply Network for Technical Competence
- ☐ **BO4** Gathering Value-adding Feedback from Projects

- ☐ **PS1** Planning Project Design Management
- ☐ **PS2** Planning Supply Chain Management Project Practice
- ☐ **PS3** Planning Integral Value Engineering Project Practice

- ☐ **PT1** Applying Design Management Practices
- ☐ **PT2** Applying Supply Chain Management to a Project
- ☐ **PT3** Selecting Supply Chain Members at the Project Level
- ☐ **PT4** Implementing Integral Value Engineering on a Project

- ☑ **PO1** Applying ADePT to Design Management
- ☐ **PO2** Applying DePlan to Design Management
- ☑ **PO3** Modelling Project Design Processes
- ☐ **PO4** Assembling Supply Chains
- ☐ **PO5** Applying Value-adding Tools to Design Problems

Identifying Matrices for Review

When ADePT is applied to an individual project, a dependency structure matrix is constructed to map out the information interdependencies of design tasks. Matrix analysis software rearranges the order of tasks within the matrix to minimise the extent of iterative information flows by locating related design tasks near to each other in the overall sequence of tasks, forming blocks of interrelated design tasks. This optimised sequencing then provides the basis for design programming. An ADePT review pulls together the matrices generated on several projects to determine if the blocks of interrelated design tasks are generic across some or all of the projects.

ADePT highlights each design task block, classifying it as either an 'A' or 'B' type depending on the strength of the information interdependencies within it (class A being stronger). Each block of design tasks is allocated a descriptive name to represent the underlying technical issues that links them.

At this stage in the review, blocks remain associated with the projects in which they occurred. You may be able to identify a number of blocks with similar content. Their occurrence in a significant proportion of the project matrices reviewed would indicate their likely generic nature - an attribute that will be confirmed (or otherwise) during the next steps of this practice.

To increase the likelihood of your review generating generic findings it is important that the projects you review:

- have a technical content typical of the projects undertaken by your organisation; and

- involve other organisations typical of those you work with in design chains.

	A	B	C	D	E	F	G	H	I	J	K	L	M	N	O	P	Q	R	S	T
Task A	■	X																		
Task B	X	■																		
Task C	X	X	■																	
Task D	X			■																
Task E		X			■								X							
Task F		X	X			■											X		X	
Task G		X		X			■													
Task H	X						X	■												
Task I				X	X		X		■											
Task J			X	X					X	■										
Task K								X			■			X						
Task L											X	■							X	
Task M								X					■		X				X	
Task N				X							X			■	X					
Task O			X								X				■					
Task P														X	X	■				
Task Q															X		■	X		
Task R																	X	■		
Task S								X											■	
Task T												X								■

Providers

- ADePT generates a matrix for each project undertaken. Depending on the project and the scope of the design process model from which it was derived (see Modelling Project Design Processes (PO3) for more information on design process models), it may represent the activity of one or more members of the value system. It is important, therefore, that each organisation ensures its design tasks are fully represented in order for its competencies to be recognised by other design chain (or value system) members. Being fully represented is best achieved by collaborating with receivers when they periodically review providers' involvement in the model.

Receivers

- Receivers must balance the number of projects reviewed and the need to provide adequate reviews that represent the full scope of the design tasks performed in typical design chains.

Identifying and Describing Design Blocks

To perform your review, you must gather the ADePT matrices from all the projects you wish to analyse. Spend some time familiarising yourself with their content before starting the review proper. Find a space for your review big enough to layout prints of these matrices for visual comparison.

Identify the first design block in the first matrix. Label this block using a short text phrase describing the common purpose of its design tasks. Construct a table for each design block found (see figure). Enter the tasks you found in the first block into the first project column in the table.

Design Block Y				
Task Description	Present in Projects			
	Project 1	Project 3	Project 4	Project 5
Block Design Task 1	X		X	
Block Design Task 2	X			
Block Design Task 3	X	X		
Block Design Task 4	X	X	X	X
Block Design Task 5	X	X	X	X
Block Design Task 6			X	
Block Design Task 7		X	X	X
Block Design Task 8	X	X	X	X
Block Design Task 9	X	X	X	X
Block Design Task 10	X	X		X
Block Design Task 11		X		
Block Design Task 12			X	
Block Design Task 13		X		X
	8	9	8	7

32 Total Observations

52 Possible Observations

$$\text{Generic ratio} = \frac{32}{52} \quad 0.62$$

Search for the same block in the next project. If you find that this next block shares the majority of its design tasks with the block content of the first project, then record a new instance of it in the table. Add any additional design tasks into the block descriptions that are present in the block currently being reviewed. Repeat this process for related blocks in the remaining project matrices. As you extend the content of each block description check that its scope remains consistent with its original title. Tasks added to a block description can change its nature to a sufficient extent so that the design tasks within it would be better described by dividing it into two smaller blocks. If this situation arises, then the two new blocks should be created.

When you have found all instances of the block, calculate its 'generic ratio' to provide a 'rule of thumb' indication of its repeatability on projects.

Compiling an IVE Opportunities Register

Each generic design block represents an opportunity to use IVE to resolve or overcome the interdependency of its consistent design tasks. Further, closely related design tasks within generic blocks represent a more focused opportunity to use IVE (and associated value-adding tools) to resolve their complexity. Within each block description (i.e. each table), search for any design tasks that are particularly closely related. Review the tables describing the content of each design block to identify any such opportunities to focus IVE. You will find that two types of opportunity arise:

1. When the technical characteristics of a project necessitate the completion of some design tasks together (unshaded shapes in figure below). These should be recorded because they may also arise on future projects sharing these technical characteristics.

2. When the technical content of the design tasks themselves causes them to be closely associated (shaded shares in figure below). These will arise on all projects where the associated generic blocks are present and contain the relevant design tasks.

Once all the block tables have been reviewed, a register of IVE opportunities should be complied. Any relevant comments should be added to the record to guide future IVE applications.

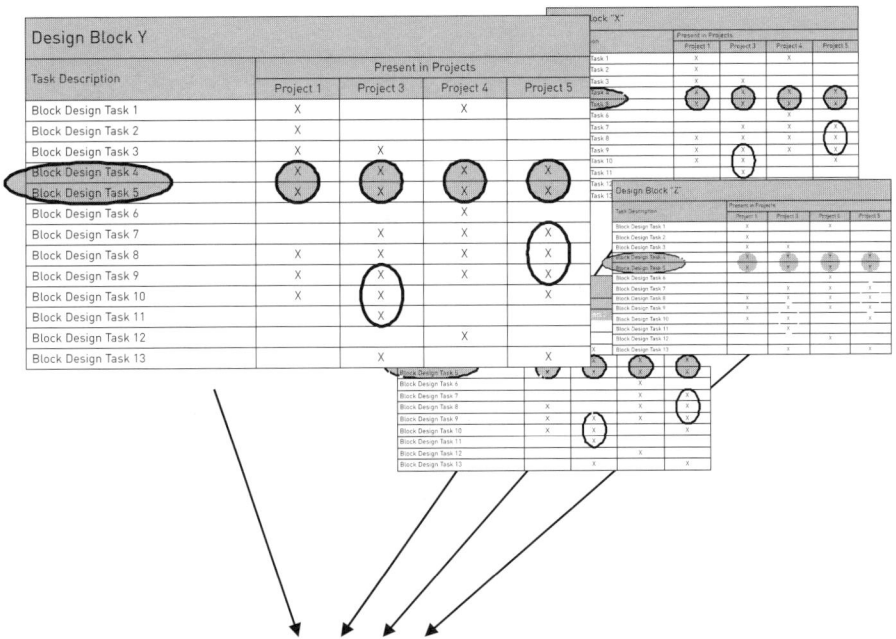

Disseminating to Future Projects

The findings should be disseminated back into the projects, for use in design management. Two types of information will be useful to projects:

1. A generic design block summary (see top figure) will help project design managers predict the elements of their design activity in which IVE will be most useful, allowing that application to be pre-emptively planned.
2. The IVE opportunity register (see lower figure) will help project design managers review their projects for the presence of design tasks that represent opportunities for more focused IVE application.

Generic Design Block Summary

Block Description	Generic Ratio	Design Disciplines Involved					Comments
		Architectural	Civil	Structural	Mechanical	Electrical	
Generic Design Block 1	0.82	X		X			
Generic Design Block 2	0.71			X			
Generic Design Block 3	0.64	X	X	X			
Generic Design Block 4	0.57	X			X	X	
Generic Design Block 5	0.44	X					

IVE Opportunity Register

Task Description	Design Disciplines Involved					Comments
	Architectural	Civil	Structural	Mechanical	Electrical	
IVE Opportunity 1	X		X			
IVE Opportunity 2		X	X			
IVE Opportunity 3				X	X	
IVE Opportunity 4	X				X	

IVE resources (such as a value-adding toolbox) can provide a mechanism to disseminate information back from the business domain into the project domain. Where the use of these resources is already established in the project domain, this will ease the introduction of the knowledge generated by this review into everyday practice.

B01

Modelling Business Design Processes

Provider benefits

- By developing a generic process framework, providers and receivers are able to establish a structured approach to design management that embodies a common language. This will allow them to align themselves both internally and externally within the supply network, and a design chain.
- Modelling design processes in detail helps providers understand the information requirements of the receivers they support. This helps them release relevant flows of information, at appropriate stages in the design process, with appropriate levels of detail.
- Modelling creates opportunities for providers to align the scope of their design with the requirements of the receivers they support by using the information requirements described in the models of their receivers' design processes.
- By informing design managers of the provision of either too much or too little design information, or at an inappropriate time in project progression, modelling will improve the efficiency of design production by providers.

Receiver benefits

- By defining their design process, receivers can coherently communicate their information requirements to supporting providers and pull information from them.
- By understanding how design information is produced as it flows back and forth between its supporting providers and itself, a receiver can optimise its design processes according to the relative importance of information flows between the receiver and providers in the design chain.
- Efficient co-ordination of information flow with providers.

Purpose

By modelling their business processes, organisations can determine their own generic process framework and extend it to include their design and project management processes in a simplified format. This will help them to align their internal functions and external interfaces by communicating their desired design scope and information requirements to other supply network members. By modelling business processes, supply network members can communicate using a common language, allowing the coherent negotiation of the interfaces between them according to their shared understanding of the flow of design information.

Summary

In order to gain a detailed understanding of their design processes, each ICD organisation must be sure where the design functions sit within the context of their business and project operations. This can be achieved by developing a high-level generic process framework of their entire business and project processes, enabling the clear definition of where design resides in relation to other activities. This provides the context for the organisation to derive a detailed understanding of their design processes, through development of a design process model that represents individual design tasks and the flow of design information between them.

When an individual supply network member develops models, internal information flows are represented and, when models are collaboratively developed within a supply network, the flow of design information between network members can be represented. Supply network members can be

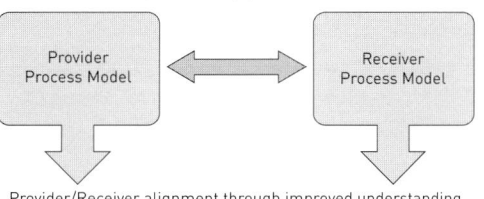

Provider/Receiver alignment through improved understanding of organisational activities and information flows

aligned according to their shared understanding of this information flow. By comparing the models developed by individual organisations in the network, any gaps or duplication in information flow can be identified. Gaps should be eliminated by allocating responsibility for the associated design tasks to appropriate network members. When design processes are duplicated, the relevant organisations should be reviewed and responsibility allocated to those with appropriate technical competencies, and who are best placed to manage the associated design risk.

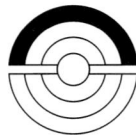

Outline Procedure

A high-level process framework can be developed by:
1. Developing a process framework
2. Applying process management
3. Maintaining the process view
4. Maintaining a legacy archive

Detailed design processes can be mapped by:
5. Defining a work breakdown structure
6. Identifying the flow of information
7. Building a graphical model
8. Validating the model.

Requirements and Resources

- Before a process model can be developed, the organisations involved must have first adopted the ICD principle of applying process management, which is required so they can define their design processes. This will help them integrate, plan and manage their design process. Practices, such as Applying Design Management Practices (PT1), will help them do this.

- Receivers are responsible for communicating their information needs to providers when developing process models. This is to ensure that the provider does not prescribe the receiver/provider relationship.

- To ensure models are accurate, the individuals involved in the process must be knowledgeable with regard to the processes involved.

- An appropriate software tool could be used to assist in the production of tabular and/or graphical views of design process models. Typical tools range from simple graphical toolkits (e.g. Visio) to fully integrated CASE tools (e.g. System Architect).

Related Practices

- ✔ **BS1** Planning Design Process Management
- ☐ **BS2** Planning Supply Chain Management Business Practice
- ☐ **BS3** Planning the Implementation of Integral Value Engineering across the Business

- ✔ **BT1** Applying Process Management in the Business
- ☐ **BT2** Aligning Supply Networks
- ☐ **BT3** Applying Supply Chain Management in the Business
- ☐ **BT4** Applying Integral Value Engineering in the Business
- ☐ **BT5** Conducting a Value Survey
- ☐ **BT6** Performing an ADePT Review

- ☐ **BO1** Modelling Business Design Processes
- ☐ **BO2** Auditing the Supply Network
- ☐ **BO3** Auditing the Supply Network for Technical Competence
- ☐ **BO4** Gathering Value-adding Feedback from Projects

- ✔ **PS1** Planning Project Design Management
- ☐ **PS2** Planning Supply Chain Management Project Practice
- ☐ **PS3** Planning Integral Value Engineering Project Practice

- ✔ **PT1** Applying Design Management Practices
- ☐ **PT2** Applying Supply Chain Management to a Project
- ☐ **PT3** Selecting Supply Chain Members at the Project Level
- ☐ **PT4** Implementing Integral Value Engineering on a Project

- ☐ **PO1** Applying ADePT to Design Management
- ☐ **PO2** Applying DePlan to Design Management
- ✔ **PO3** Modelling Project Design Processes
- ☐ **PO4** Assembling Supply Chains
- ☐ **PO5** Applying Value-adding Tools to Design Problems

Developing a Process Framework

Organisations should review existing process frameworks (such as the RIBA *Plan of Work*[1] and the Process Protocol[2]) and determine the generic definitions of the business and project processes associated with their operations, and compare these with those currently undertaken by the organisation. Because a process framework map is generic it should encompass the full breadth of processes that a business may perform in delivering products and/or services to its customers. The processes actually performed by the business are overlaid, or inserted, into this generic process framework, creating an initial definition of the business processes, which can then be customised when applied to specific projects or commissions. The Process Protocol has been designed specifically as a flexible framework that can be modified to suit specific types of organisation, project or construction.

Providers

- Providers can develop their own process framework, which they can then customise to meet the needs of their role(s) in 'projects' in which they are involved. Alternatively, they can contribute to the development of a process framework by the receiver they support in a supply network (or value system). Their role in individual projects will then be defined by the manner by which that receiver customises a generic framework for specific project applications.

Receivers

- Receivers should adopt a consistent process framework to provide a basis for working with a supply network.
- When the receiver forms the primary link between individual project design chains and project customers, they should identify whether these customers use particular process frameworks to manage their projects. A framework used by several customers may influence the development of design frameworks in the supply network.

Applying Process Management

Process management develops and operationalises the process framework and is responsible for planning and monitoring each phase. It should include maintenance, review and audit activity to provide structured feedback that will enhance business operations and to identify additional training needs. A single person or authority within an organisation should oversee its process management to provide a consistent, accountable approach to the maintenance, dissemination and diffusion of the framework. They also should be independent of projects and take a change of management role to actively promote the uptake and application of the framework among designers and managers.

---

[1] RIBA (2000) *The Architect's Plan of Work for the Procurement of Feasibility Studies, a Fully Designed Building Project, Employer's Requirements of Contractor's Proposals*, RIBA Publications, London.

[2] See Figure 2.8 and www.processprotocol.com.

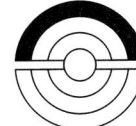

Maintaining the Process View

The generic process framework must have the flexibility to enable the network (or individual organisation) to align itself with the particular commercial circumstances of individual projects. While organisations that repeatedly work together in similarly-structured design chains may seek to work together in an integrated fashion, there is no reason why individual organisations can revert to more traditional contract relationships (where the rules of engagement are known) for individual projects.

When project activity identifies faults and improvements to the business process framework, these insights must be fed back from the project domain to the business domain so that the benefits can be passed to other live projects, and (through revised process frameworks and design process models) to future projects.

Providers

- Providers should ensure that any changes in their competencies and project responsibilities are reflected in the revisions of the generic process framework and detailed level design process models for application to subsequent projects. This will ensure that their model continuously reflects their current abilities and is particularly important if they are in the process of extending, or otherwise changing, the scope of the services that they offer within a supply network.

Receivers

- Receivers must ensure that their generic process framework continues to contain processes that allow them to meet the needs of their clients. Client (or end-user) satisfaction with the service they provided must, therefore, form part of the assessment made at the end of each project to determine how well the framework has performed and how the generic framework should be modified for application to subsequent projects.

Maintaining a Legacy Archive

When using a generic process framework, it is important for the organisation to establish a legacy archive, which contains all the documentation produced during the progression of each of the projects to which it has been applied. This archive may be most readily established if linked to, or built upon, the existing document management systems used by the organisation. The information in this archive, in addition to describing the technical solutions and working methods deployed in each project undertaken by the organisation, also describes the manner by which the generic process framework definitions were modified for application to individual projects. This information should be periodically reviewed by an organisation to help it learn and develop its process management competencies.

Providers

- If the legacy archive is shared among supply network (or value system) members between projects, then this will allow the learning obtained through the content of the archive to be sustained across projects. Organisations that were not involved in a particular project undertaken by the supply network or value system can learn from the experiences entered into the legacy archive.

- A shared legacy archive helps providers understand the solutions to technical and process problems that have been found effective in the past. This should encourage their development and adoption on subsequent projects.

Receivers

- By feeding knowledge of how processes were undertaken back to supporting providers, they can better understand and improve their project and design management roles. This, in turn, will enhance their reputation with providers and, hence, future working relationships.

Defining a Work Breakdown Structure

To model the relationship of design tasks, each design task must first be identified. An effective way of doing this is by defining a work breakdown structure to represent the full scope of the design competencies of the organisation (irrespective of whether they are required on a particular project). This creates an opportunity for design chain members to communicate their design capabilities to others. To ensure accuracy, a work breakdown structure must be developed from the bottom up by capturing the knowledge and expertise of designers and managers. The structure may be assembled by collecting information on the different technical systems the design chain member typically designs, as this helps structure the identification of the detailed level design processes undertaken within each system. The lowest level tasks are often associated with the production of design deliverables (e.g. drawings, specifications and the supporting calculations).

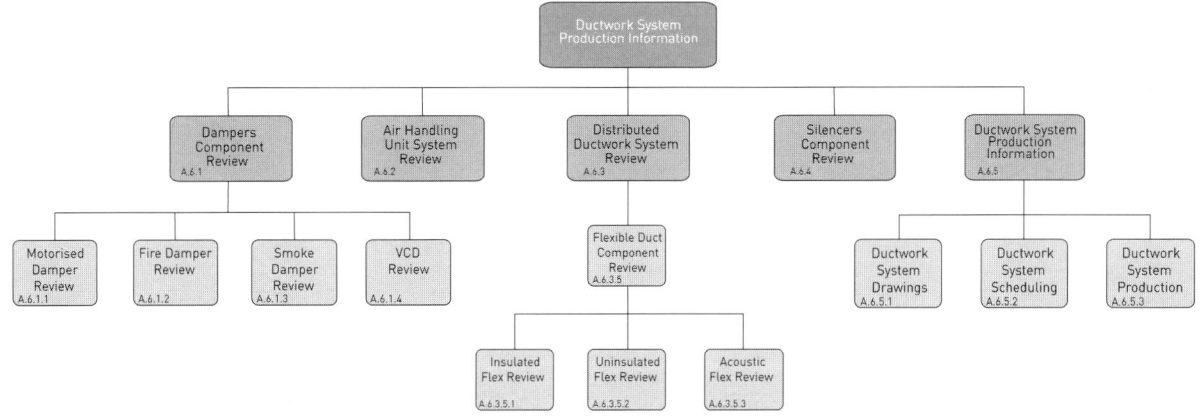

It is likely that most businesses will already have well-defined work breakdown structures, although the processes may not be described as such. Process management in the business domain will help to refine these work breakdown structures and to develop a coherent set of work breakdown structures for members of the supply network.

Providers

- The provider must clearly define all the systems that they can offer as part of their design capability, including all sub-systems and detailed level design processes. A workshop environment provides an excellent forum in which to capture this data from the experienced knowledge keepers residing within their organisation.

Receivers

- Different providers in the same or similar fields of operations are likely to have different approaches to work breakdown structures and/or process management. When a receiver co-ordinates the design process frameworks of a number of such providers, care must be taken as to which process information is generic and which is confidential (because the latter gives a competitive advantage between providers).

Identifying the Flow of Information

To determine how design processes are related to each other, you must define the information flowing between them. Start by considering the design information flow within your organisation before modelling the 'cross-organisational' flows of information that link design processes between multiple organisations. The process modeller should brief designers and capture the required information through the annotation of activity lists or diagrams. Ask designers to concentrate on defining their information requirements (i.e. the inputs), which they can easily identify. Care must be taken to distinguish between their information requirements and the project deliverables that they produce using that information. *Co-ordinated Project Information* provides guidance on common deliverables, which may assist in this differentiation[3]. The process modeller can then 'connect' these information inputs to the appropriate tasks generating them (as outputs). Some, but by no means all, of these outputs will be deliverables.

Providers

- By mapping the information flow between activities, providers will develop a fuller understanding of the complex process in which they engage.

Receivers

- Receivers should seeks to understand the information needs of providers to avoid information overload and to ensure design management is focused on critical information not just on generic deliverables.

Building a Graphical Model

The amount of data generated by the work breakdown structure of activities and the identification of the flow of information can be large and complex. It is therefore advantageous to represent this data in a graphical form. These models will illustrate the business design processes in a simple manner, making them easier to validate and apply. A suitable notation (e.g. IDEF0) should be used to ensure that the process is represented in a clear and consistent manner. Preparation of the model will also facilitate the verification of the data (i.e. the integrity check that ensures full consistency of the model) - with no loose ends.

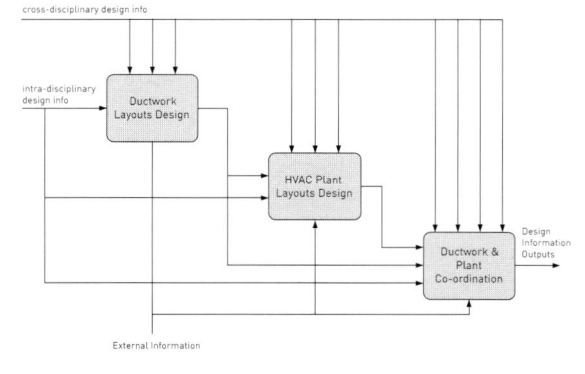

Validating the Model

An organisation can only use its design process model effectively if it knows that it accurately represents its processes. Before using a model, therefore, it must be validated. The model is reviewed by the organisation's designers to confirm that it accurately represents the operations of the organisation, and that the flow of design information across its interfaces with other design chain members is accurately represented. This is best achieved through a series of review workshops. To be effective, the key individuals responsible for managing the organisation's design processes must attend these workshops. The process modeller should then update and distribute the validated model.

[3] Co-ordinating Committee for Project Information (1987) *Common Arrangement of Work Sections for Building Works*, Building Project Information Committee, London.

Auditing the Supply Network

Provider benefits

- The supply network audit helps providers understand their SCM competency and to compare it to other members of the supply networks.

- The audit gives providers access to a database of those SCM practices used throughout the supply network, allowing the providers to make their own assessment of their SCM maturity.

Purpose

To establish the supply chain management (SCM) competencies present within a supply network and to identify, for each member, the practices that give rise to these competencies. By auditing a supply network, the audit outcome can be used by network members to understand where they lie in relation to other network members. The audit also provides useful insights into how supply chains and design chains can be assembled from the network.

Receiver benefits

- A detailed understanding of how the necessary SCM practices are distributed among supply network members and to what level of maturity.

Summary

Receivers can use this practice to structure their assessment of a provider's SCM maturity and competencies. The audit results in a simple classification of the SCM maturity of each surveyed organisation, by observing the use of various practices within the organisation (see figure below). The audit also generates some of the information providers and receivers need to align their functions within a supply network (the remainder of the information comes from Modelling Business Design Processes (B01)). This information is also useful when establishing supply chains. The audit applies the process described in Applying Supply Chain Management in the Business (BT3) to other organisations to help them benchmark their respective roles in the supply network. This assessment is structured around a matrix called the SCM Maturity Matrix, which creates a series of cells by allocating levels of maturity against seven key aspects of SCM. The five levels of maturity are described below:

Level	Name	Definition
5	Mature Implementation	SCM is fully embedded and practised on all projects throughout the organisation. The organisation is no longer conscious of its use of SCM practises, but views them as part of its normal practice.
4	Full Implementation	Organisation-wide implementation of SCM practices that have been validated from isolated experiences of their use at lower maturity levels.
3	Partial Implementation	The organisation has made the business decision to adopt SCM, based on its tentative implementation experiences. Senior management supports the piloting of SCM practices where greatest benefits are anticipated.
2	Tentative Implementation	Individuals drive SCM practice on isolated projects. Individual project managers are trying out a few SCM practices on their own initiative using their own projects and relying on their personal relationships to do so.
1	Awareness of SCM	The relevance of SCM to the organisation has been recognised by key individuals at all organisational levels, including senior management.
0	No SCM	No evidence of SCM is present in the organisation. A variety of traditional practices are used throughout the business to perform various functions.

Outline Procedure

A supply network can be audited by:
1. Initiating the audit
2. Assembling provisional supply network
3. Administering the SCM audit
4. Generating a network profile
5. Planning SCM business development

Requirements and Resources

- Organisations will have needed to have performed Planning Supply Chain Management Business Practice (BS2) and Applying Supply Chain Management in the Business (BT3) prior to commencing the auditing process. Auditing the Supply Network for Technical Competence (BO3) can be carried out as a pre-qualification to auditing the supply network to ensure that a minimum standard of project performance exists before assessment for membership of a supply network.

- Involvement of senior management is required on the part of receivers and providers during the audit process, however, for receivers, after the initial assessment and compilation of the SCM Maturity Matrix, the administration and calculations for the audit should not require the involvement of either specialised staff or senior management

Related Practices

- [ ] **BS1** Planning Design Process Management
- [✔] **BS2** Planning Supply Chain Management Business Practice
- [ ] **BS3** Planning the Implementation of Integral Value Engineering across the Business

- [ ] **BT1** Applying Process Management in the Business
- [ ] **BT2** Aligning Supply Networks
- [✔] **BT3** Applying Supply Chain Management in the Business
- [ ] **BT4** Applying Integral Value Engineering in the Business
- [ ] **BT5** Conducting a Value Survey
- [ ] **BT6** Performing an ADePT Review

- [✔] **BO1** Modelling Business Design Processes
- [ ] **BO2** Auditing the Supply Network
- [✔] **BO3** Auditing the Supply Network for Technical Competence
- [ ] **BO4** Gathering Value-adding Feedback from Projects

- [ ] **PS1** Planning Project Design Management
- [ ] **PS2** Planning Supply Chain Management Project Practice
- [ ] **PS3** Planning Integral Value Engineering Project Practice

- [ ] **PT1** Applying Design Management Practices
- [ ] **PT2** Applying Supply Chain Management to a Project
- [ ] **PT3** Selecting Supply Chain Members at the Project Level
- [✔] **PT4** Implementing Integral Value Engineering on a Project

- [ ] **PO1** Applying ADePT to Design Management
- [ ] **PO2** Applying DePlan to Design Management
- [ ] **PO3** Modelling Project Design Processes
- [ ] **PO4** Assembling Supply Chains
- [ ] **PO5** Applying Value-adding Tools to Design Problems

Initiating the Audit

When used for the first time, the content of the maturity matrix (which is used to structure a supply network audit, see Figure 1) will be generic and identify the practices (Figure 2 shows how the practices associated with the key aspects are documented) sought from network members. This can be used to select potential providers for network membership if a receiver seeks to establish such a network. When used later to audit the same supply network, it provides an opportunity to re-align the SCM competencies of the supply network members to the business needs of the receiver who is undertaking the audit. These business needs will have been identified in Planning Supply Chain Management Business Practice (BS2). The audit is conducted in two main stages:

- The first step is for the receiver to review its strategy for integrated SCM practices into its business plan in order for the receiver to be clear on which SCM practices it needs.

- The next step is to update the practices within the SCM Maturity Matrix, used to structure these assessments, to ensure that they will be able to support the receiver's current, and anticipated future needs.

This two-stage process is conducted in workshops made up of reviewers occupying strategic roles from the receiver organisation.

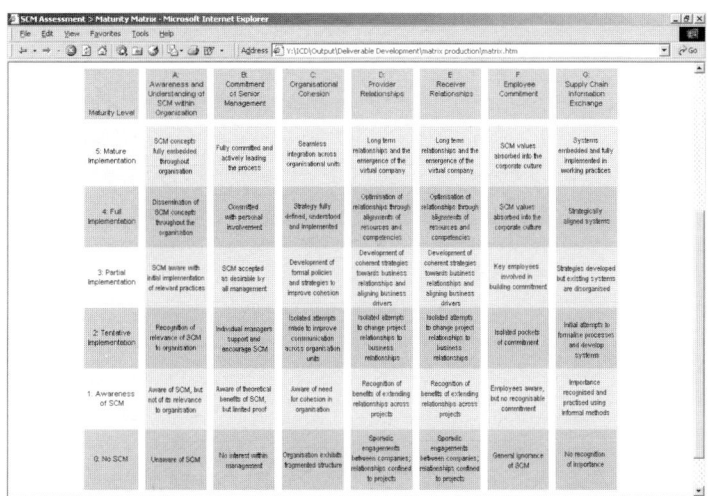

Figure 1

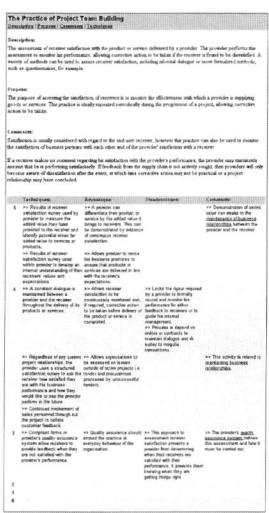

Figure 2

Receivers

- Receivers select a supply network on the basis of their needs. The selection is a product of the receiver Planning Supply Chain Management Business Practice (BS2).

- Auditors for the supply network audit should be drawn from a variety of areas within the business. Care should be taken to ensure that the process is not monopolised by a single function or department. It is recommended that the process involves senior management to provide the necessary strategic input, together with project personnel to ensure the audit assesses SCM practices relevant to the needs of the business's current projects.

- Work in developing the network needs to be placed in the context of where the business sees itself and its likely future direction. The supply network is a mechanism to promote and apply SCM and so attention must also be paid to the perceived role of SCM to other supply network members who may have divergent views as to its relevance and suitability for their own organisation's future development.

- The SCM audit is likely to be used to inform an organisation's SCM development strategy. When the audit is repeated, the practice content of the survey matrix should be amended to ensure that the audit searches for the presence of SCM practices that are appropriate to the surveying organisation's business needs.

- During this stage, the receiver has the opportunity to identify those practices it considers important for the supply network to be using.

Assembling Provisional Supply Network

As discussed above, the audit, when first used, can be used to set a benchmark for initial supply network membership. This is represented by the need for a surveyed provider to practice SCM in the manner (i.e. using the practices) identified in the current level of SCM maturity. The first iteration of the network will have a provisional membership because member performance will not have been reviewed. After a period of SCM activity and an anticipated increase in the proficiency with which SCM practices are used within the supply network, its initial quality will be assured, as each member will have passed the initial benchmark. When used subsequently, audits are an opportunity to:

- match business needs between receivers and network members;
- identify gaps in the network membership (i.e. identify any practices sought in the maturity matrix but which are not performed by any network members (or are not performed with the desired level of competency)); and
- to assess the performance of existing members.

Two activities are involved in assembling a provisional network:

- Selecting providers for an initial audit. The selection will be based on the providers' prior performance, geographical spread and financial resources.
- With the initial selection in place, potential providers are audited via workshops as discussed above.

Providers

- The nature of the audit may differ depending on whether the receiver intends it to be carried out in a facilitated workshop or through a self-assessment process.
- The receiver may prepare the ground for the audit in a variety of ways. The audit may be introduced using newsletters, 'supplier suppers', road-show presentations, or on an individual business to business basis. In addition, standard presentations, with supporting documentation, may also be issued to provide background information and to describe for providers how they can fit in to the process and how the process relates to the receiver's business objectives.

Receivers

- SCM competency is not the sole determinant of supply network membership. SCM competency can only be used in conjunction with other assessments of provider technical competencies; these might include assessments of CAD or 3D modelling capability or of specific technical knowledge.
- Due to constraints an organisation faces on its capacity to deliver design solutions, such as resource availability or workload, there may be some overlap between SCM competencies of the supply network and those of the receiver where specific design tasks can be equally well performed by either the receiver or a provider from the supply network. This situation can provide some useful flexibility on projects, allowing activities performed by receivers and the network members to be varied depending on market or project circumstances.

Administering the SCM Audit

The SCM audit allows an SCM maturity profile to be generated for an organisation. This can be used to inform the organisation whether it has the SCM competencies it needs to achieve its strategic objectives. The process of deriving this profile is normally performed in a workshop, which should be attended by the individuals representing a cross-section of the provider's business.

A provider may use its SCM audit profile to benchmark itself against the practices of other providers in the supply network. This can be done through subjective or evidence-based assessments or through a formal external audit. This will help the provider deploy its own business strategies as its performance, relevant to its partners within the network, can be monitored. The audit compares seven key aspects of SCM practice, deriving a 'profile' of SCM maturity for each audited organisation (see figure).

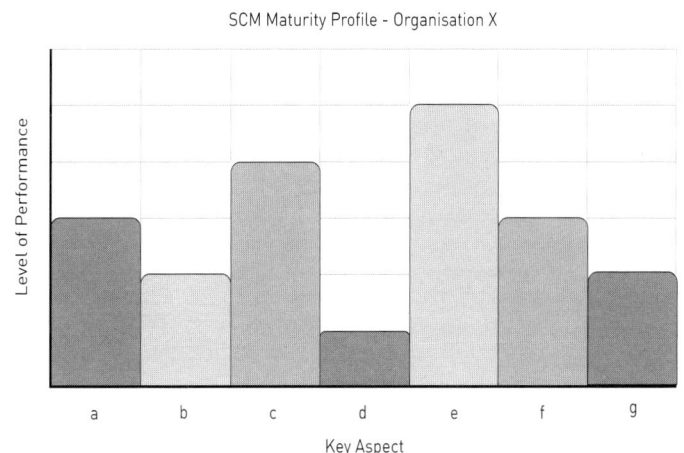

SCM Maturity Profile - Organisation X

Providers

- Those individuals who will provide input should be drawn from a variety of areas within the business. It is recommended that the audit involve senior management to provide the necessary strategic input, as well as project personnel to represent the organisation's operational SCM activities.

- The audit workshop will be more productive if the information to be assessed is gathered in advance. This might include such things as case studies or procedure manuals.

- The desired level of SCM maturity should be at a point that reflects the normal working practices of most people at the workshop. It is always likely that, on an isolated project where other practices are used, some organisations will get different SCM maturity results, but it should be recognised that such projects do not usually provide accurate SCM maturity indicators for the business as a whole.

- The provider has an opportunity to present its capability with an auditing receiver by proposing other valid practices and supporting techniques considered to increase its maturity level.

Receivers

- Receivers should apply the SCM audit to their own organisations to provide a comparison for the providers in the supply network.

Generating a Network Profile

The generating network profile stage brings together the results of each organisation's self-assessment. Generating a profile comprises two activities, after which the resulting profile is fed back to the audited providers:

1 The first step is to review the SCM Maturity Matrix in the light of additional practices that might have been identified by the audit. This review provides an opportunity to update the matrix to make it a more effective shared knowledge repository for the network.

2 The second step is to calculate provider SCM maturities for each key aspect for the network and from this an SCM Maturity Index for individual audited providers (see figure). The former summarises a provider's current SCM maturity for each key aspect against the reviewed SCM Maturity Matrix and allows direct comparison between supply network members. The index provides a single score of an organisation's position relative to others in the network, while maintaining confidentiality over actual maturity scores achieved by an organisation for each key aspect.

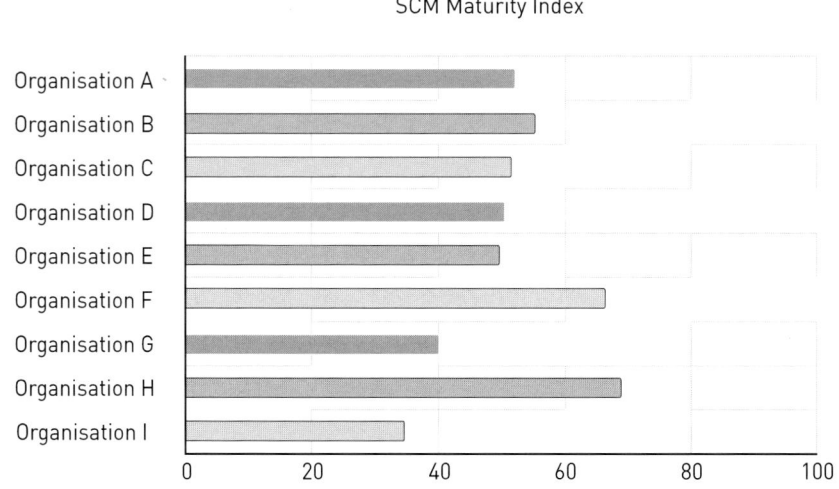

SCM Maturity Index

Providers

- To perform a self-assessment, each organisation must identify documented evidence of its use of the practices it is claiming to use in the assessment if the use of an SCM practice would give rise to such documentation. In the case of providers, this evidence may include project case studies or extracts from practice manuals, for example.

- The provider is given a revised SCM Maturity Matrix. The provider then has the opportunity to review the matrix in case it wishes to make alterations to its SCM development plan, to reflect either the possible changes to the matrix, or, because the provider will now know its comparative maturity rating, to alter the trajectory of the development plan. To aid this, a large supply network may be sub-divided into technical specialisms.

Receivers

- The review of practices should not necessarily be limited to a desk study. The opportunity exists to visit providers who have identified new practices or new ways of supporting practices and to learn more and develop a deeper understanding of the new practice and the provider's ways of working.

- Desired practices are those that the receiver deems valid for the network. The revised SCM Maturity Matrix is fed back to the members of the supply network and becomes the basis of the next iteration of the audit.

- It is not recommended that members receive copies of each other's profiles until an adequate level of trust has developed among the network members. Organisations need to have reached the point where they are prepared to learn openly from one another before they can really profit from seeing each other's profile.

Planning SCM Business Development

Planning SCM business development uses the SCM audit findings as the basis of an organisation's SCM development and implementation strategy. The revised SCM Maturity Matrix can identify tools and techniques that an organisation may wish to consider adopting in order to develop its SCM capabilities. Organisations can identify any problematic differences between the network's SCM maturity profile and that of any individual organisation. Organisations can use this comparison to plan their strategic SCM development by:

- working together with appropriate members of provider organisations to confirm that discrepancies between the organisation's and the network's SCM maturity profile are inappropriate and require strategic action;
- targeting practices to bring the maturity of problematic key aspects to the required level;
- establishing strategies and plans to improve existing practices to support these key aspects;
- identifying any new practices that may also be used to improve key aspect maturity; and
- establishing and implementing strategies and plans for improvement and using, if possible, pre-defined strategic and project-level implementation plans.

Receivers

- The SCM audit process is concerned with improving the SCM competencies of providers and receivers. However, it is recognised that the receiver may consciously wish to accept lower levels of maturity in certain areas in the supporting supply network in order to ensure that all organisations are capable of operating at the same level of competence rather than allow the supply network to fragment into smaller groups of organisations of similar maturity. The audit may be used to inform the strategic development of the receiver's SCM competencies in the same way as it can help providers with their strategic development.

B03

Auditing the Supply Network for Technical Competence

Provider benefits

- For the provider, it is often easier to identify the technical skills that need to be updated or acquired than it is to understand any culturalor organisational change that might be needed. The new management skills or cultural change that a company might need can be found by conducting a supply network audit (details of how to do this are given in Auditing the Supply Network (B02)). When it is clear that the level of technical competency needs to be raised, the act of improvement can also be a vehicle for introducing cultural or organisational change.

Receiver benefits

- Structured interviews answer the need for a visibly rigorous standardised assessment process. The questions asked and the matrices of standard answers can be easily customised to suit the individual needs of the supply network.
- Building upon existing assessment methods, such as that given in *Selecting Contractors by Value*,[2] provides an assessment process that is both accessible and robust, making it possible to introduce it easily into both the assessing (the receiver) and the assessed organisation (the provider).

Purpose

This practice is used to establish the technical capabilities that exist within the supply network, or within organisations being considered for inclusion in the supply network. By benchmarking the technical competencies of organisations, not only is there a basis for selecting which companies should work on which projects, but minimum standards can be established for the network and these can then be used to measure improvement.

Summary

This practice details a number of techniques that can be used to measure the level of technical competency in a network. This is essential information, because when it is known which skills exist and which are needed, the network (usually the receiver) is then in a position to increase the skills it possess (by providing feedback to providers) and to acquire the ones it lacks. This process can also be useful for the individual providers, because not only will all the providers know the value of what they contribute, and what they need to do to increase their value to the network, but they will understand what each other is capable of, and, thus, be in a better position to trust the people with whom they are collaborating.

The technical audit can be made up from any one or combination of methods, including:

- **Structured Interviews** - where the receiver interviews key personnel from provider companies.
- **Facility Visits** - where the receiver conducts visits to the places where the providers work, such as design offices and manufacturing facilities, to see what equipment, resources and innovative methods are being used.
- **Key Performance Indicators (KPIs)** - these are the precursor to enable companies to benchmark themselves and others. A range of KPIs for the construction industry have been compiled by the Construction Best Practice Programme[1] among others, which can be used without firms having to develop their own. These give a common measure of capability and a means for assessing performance.
- **Pre-qualification Questionnaires** - questionnaires designed to elicit basic information about organisations, traditionally used on a project by project basis to build tender lists.

[1] A government-led industry specific programme established to raise awareness of the benefits of best practice and to provide guidance and advice to organisations in the UK construction industry in order to ensure they have the knowledge and skills necessary to implement change. See www.cbpp.org.uk.
[2] Jackson-Robbins, A. (1998) *Selecting Contractors by Value*, CIRIA, London.

Outline Procedure

There is no set order to the methods discussed in this practice. Individual organisations should seek to select and use techniques that they feel are appropriate to their circumstances.

- Structured Interviews
- Key Performance Indicators (KPIs)
- Facility Visits
- Pre-qualification Questionnaires

Requirements and Resources

This practice is about establishing and maintaining a supply network. It works best within the context of ICD (and needs to be seen in conjunction with the practices on Auditing the Supply Network (BO2) and Applying Supply Chain Management in the Business (BT3)) but is not confined to ICD. If it is applied to an organisation where there is already some commitment to SCM, such as the use of pre-qualification questionnaires, then it can be easily introduced without (immediately) needing to include resource intensive exercises, such as facility visits.

Related Practices

- ☐ **BS1** Planning Design Process Management
- ☐ **BS2** Planning Supply Chain Management Business Practice
- ☐ **BS3** Planning the Implementation of Integral Value Engineering across the Business

- ☐ **BT1** Applying Process Management in the Business
- ☐ **BT2** Aligning Supply Networks
- ☑ **BT3** Applying Supply Chain Management in the Business
- ☐ **BT4** Applying Integral Value Engineering in the Business
- ☐ **BT5** Conducting a Value Survey
- ☐ **BT6** Performing an ADePT Review

- ☐ **BO1** Modelling Business Design Processes
- ☑ **BO2** Auditing the Supply Network
- ☐ **BO3** Auditing the Supply Network for Technical Competence
- ☐ **BO4** Gathering Value-adding Feedback from Projects

- ☐ **PS1** Planning Project Design Management
- ☐ **PS2** Planning Supply Chain Management Project Practice
- ☐ **PS3** Planning Integral Value Engineering Project Practice

- ☐ **PT1** Applying Design Management Practices
- ☐ **PT2** Applying Supply Chain Management to a Project
- ☐ **PT3** Selecting Supply Chain Members at the Project Level
- ☐ **PT4** Implementing Integral Value Engineering on a Project

- ☐ **PO1** Applying ADePT to Design Management
- ☐ **PO2** Applying DePlan to Design Management
- ☐ **PO3** Modelling Project Design Processes
- ☐ **PO4** Assembling Supply Chains
- ☐ **PO5** Applying Value-adding Tools to Design Problems

Structured Interviews

The structured interview is a method for carrying out a rigorous and repeatable assessment of a provider's technical capabilities. It comprises a set of high-level questions, with related sub-questions. They are asked in a standard order, separately to several employees of the same organisation in individual interviews. The structured interviews consist of two stages:

1. The actual interview.
2. Provision and analysis of supporting evidence.

The purpose of the interview is examine an organisation in order to identify if there are important differences and/or similarities in the way that it goes about its work - such as whether the management philosophy is consistently applied across the organisation. The interview requires a number of employees (at least three) to be interviewed separately. For the purposes of the supply network, these might typically be a managing director/senior manager, design manager/designer and a site engineer/designer.

The example interview sheets shown below contain a series of high-level questions with a summary of possible responses presented as a matrix. A series of subsidiary questions follow on from the high-level questions. These are designed to tease out more detailed responses to give added depth to the matrix. Part of the value of this sort of exercise is that it seeks to give a standardised assessment. It means that all interviewees need to be asked the same questions and that the interviewer is not at liberty to pursue a particular line of inquiry with follow-up questions.

The scoring of each question is based on the interviewer's judgement as he or she seeks to allocate the response to a cell of the matrix, and only the answers that are given during the interview are used when making the judgement. In this case, the criteria are capability, experience and compatibility. However, there is a second stage to the process, where evidence (such as procedures or other documentation) can be weighted to reflect the needs of the assessing organisation and generate a final score for the supplier. This weighting would be set in advance during the formulation of the assessment questionnaire. Evidence is seen to be an important factor in differentiating between providers compared to the lack of subtlety provided by the matrix.

Question: What areas of in-house design is your organisation capable of?

	Poor	Adequate	Good	Excellent	Comments
Capability					
Experience					
Evidence					
Compatibility					

		Poor	Adequate	Good	Excellent	Comments
	Capability	No in-house design capability	Design and co-ordination capability	Capable of designing systems from a detailed brief	Capable of producing design solutions in-house from concept at initial project stages	
	Experience	No experience of in-house design	Limited in-house design experience	Experience of repeatable in-house design projects	Experience of a wide range of in-house design project types in different market sectors	
	Evidence	No evidence of in-house design	Evidence of one in-house design project	Evidence of two or more in-house design projects	Evidence of several in-house design projects	
	Compatibility	No in-house IT systems for design	Some similar in-house IT systems to AMEC for design identified	Compatible in-house IT systems for design identified	Compatible IT systems for in-house design identified, with clear and established procedures developed to enable the transfer of design information to the whole project team	

Providers

- For providers, there is an overlap between the sort of information they need to present in supply network audits (see Auditing the Supply Network (B02)) and that information they need to give as evidence in structured interviews. Therefore, it can be useful, and can offer a saving in time, to undertake both exercises together.

Receivers

- As discussed above, because the structured interview is designed to give standardised objective results, there is not the opportunity for the receiver to ask supplementary questions on any area that is of particular interest. For this reason, receivers may wish to combine the interview with less rigid methods, such as the facility visit, to get a fuller assessment of the organisation

- Training of assessors in evidence-based assessment techniques helps to align the assessment across the receiver's staff.

Key Performance Indicators (KPIs)

KPIs are measures that can be used as a basis to benchmark organisations against one another. The Construction Best Practice Programme currently publishes a series of construction industry KPIs with other more specialised KPIs produced by CIRIA, ACE, ICE and others. In the context of ICD, these are applicable to both the project domain and the business domain.

On the whole, however, KPIs are a blunt instrument when it comes to building supply networks. A small number have a technical focus and are quite useful. The majority address the profitability of the organisation and are not particularly relevant to understanding supply networks. Where KPIs are valuable, however, is in providing a rough and ready measure of how the members of the supply network compare to other organisations in the construction industry. Supply networks may well benefit from devising and using a specific set of KPIs to monitor joint progress and performance.

Providers

- KPIs used can form fast and cheap measurements that generate an unambiguous score. They can also be useful in helping to breakdown some of the barriers that prevent innovation and change in organisations. This is especially true if some of the more common and widely quoted KPIs are given as the reason for why change must take place.

Receivers

- KPIs used can form fast and cheap measurements that generate an unambiguous score. They can also be useful in helping to breakdown some of the barriers that prevent innovation and change in organisations. This is especially true if some of the more common and widely quoted KPIs are given as the reason for why change must take place.

Facility Visits

This involves the receiver sending people to sites where the providers do their work, to assess how the work takes place and to see how technically competent it is. Such visits are important opportunities for the receiver to form a more rounded impression of the provider. Without them, the receiver's assessment of a provider is limited to how the provider has performed during formal/semi-formal encounters at site meetings or on ad-hoc and unstructured visits to the receiver's facilities.

The drawback to the facility visit is the time it requires from both the receiver and provider. A facility visit may involve a range of staff representing the particular specialisms that need to be assessed. Consequently, they can be more easily justified by the receiver if they can be combined with other assessment methods, such pre-qualification questionnaires or structured interviews or restricted to key members of a supply network (e.g. those whose activities are mainly performed offsite).

Providers

- Providers have few opportunities to tell a receiver about their range of competencies and receivers will often have an inaccurate and out-of-date picture of a provider's resources and the sophistication of its operations.

Receivers

- KPIs used can form fast and cheap measurements that generate an unambiguous score. They can also be useful in helping to breakdown some of the barriers that prevent innovation and change in organisations. This is especially true if some of the more common and widely quoted KPIs are given as the reason for why change must take place.

Pre-Qualification Questionnaires

The use of pre-qualification questionnaires is common within the construction industry where traditional project procurement methods are used. Their purpose is to ensure that those firms, who are asked to tender, are suitable. They are used in a similar way in supply networks - to ensure that an organisation has a minimum standard of technical competence before it is invited to join a supply network.

Typical examples are given below as to the sort of areas that are assessed with pre-qualification questionnaires. These include health and safety, environmental issues, quality assurance and insurance arrangements. However, they are deterministic in nature, making subsequent analysis and comparison between organisations difficult.

Pre-qualification questionnaires are typically given to providers to fill in and then return to receivers. They are only useful in collecting fairly basic information, such as proof of insurance, and really need to be used in conjunction with the more detailed methods discussed - facility visits or structured interviews - if a receiver wants to be able to distinguish the adequate from the excellent.

Providers

- Like KPIs, pre-qualification questionnaires have the advantage of being readily familiar to staff and, therefore, the compliance process is undemanding.

- Questionnaires can be time consuming to complete, as different receivers will have different questionnaires and may require compliance with the assessment process on a project by project basis.

Receivers

- The use of pre-qualification questionnaires has the benefit of using existing data collection methods and incorporating them into the ICD practice. Adapting the pre-qualification form from a project-based process to a business domain one means that a stage in the procurement process can be omitted with savings in time and, for the receiver, in transaction costs (the cost of going through the appointment process) as well.

B04

Gathering Value-adding Feedback from Projects

Provider benefits

- Gathering feedback creates opportunities to document the role played by providers in arriving at design solutions. Innovative solutions can be replicated on future projects and increase the likelihood of the provider being involved on future projects.

Purpose

Feedback is collected on the way value-adding tools have been used on projects to refine or to introduce new tools for design chain members. These improvements are disseminated through the value-adding toolbox for application to the remainder of the current project and future projects undertaken by the organisation or value system.

Summary

The practice captures the way a design chain works with the tools it has at its disposal. It entails monitoring and reviewing applications and feeding back to the business the IVE lessons learnt on projects. Therefore, it also helps demonstrate to the client the value-for-money from the design process. The feedback creates an audit trial of design decisions, that records the value of individual design contributions and that demonstrate where and how value has been created (see Planning Integral Value Engineering Project Practice (PS3) at the outset of the project).

Receiver benefits

- If the value-adding toolbox is operated as a shared resource within a design chain, then providers and receivers benefit from access to the collective experience of other chain members.

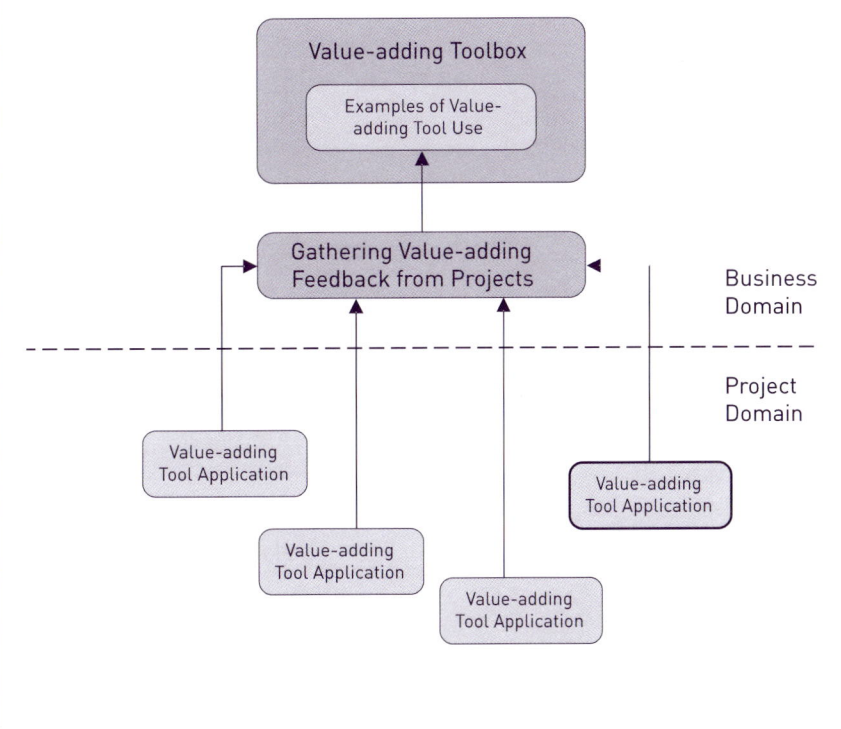

Outline Procedure

Feedback on value-adding tool use in projects can be gathered by:

1. Identifying and reviewing applications
2. Capturing informative examples (case histories)
3. Disseminating case histories

Requirements and Resources

- The toolbox is of most value when it is available in the business domains of all design chain members.
- Demonstrating value-for-money design to clients can only be achieved if designers understand its importance and actively work to achieve it.
- The IT software and infrastructure through which a value-adding toolbox is disseminated within an organisation, value system, or project, can also provide the mechanisms through which feedback on value-adding tools use is gathered. Therefore, it is essential to ensure that this mechanism is available to all individuals using value-adding tools in their work.
- Documented case histories of applications, which provide designers with valuable examples and feedback, must be available through the value-adding toolbox.

Related Practices

- ☐ **BS1** Planning Design Process Management
- ☐ **BS2** Planning Supply Chain Management Business Practice
- ☐ **BS3** Planning the Implementation of Integral Value Engineering across the Business

- ☐ **BT1** Applying Process Management in the Business
- ☐ **BT2** Aligning Supply Networks
- ☐ **BT3** Applying Supply Chain Management in the Business
- ☐ **BT4** Applying Integral Value Engineering in the Business
- ☐ **BT5** Conducting a Value Survey
- ☐ **BT6** Performing an ADePT Review

- ☐ **BO1** Modelling Business Design Processes
- ☐ **BO2** Auditing the Supply Network
- ☐ **BO3** Auditing the Supply Network for Technical Competence
- ☐ **BO4** Gathering Value-adding Feedback from Projects

- ☐ **PS1** Planning Project Design Management
- ☐ **PS2** Planning Supply Chain Management Project Practice
- ✔ **PS3** Planning Integral Value Engineering Project Practice

- ☐ **PT1** Applying Design Management Practices
- ☐ **PT2** Applying Supply Chain Management to a Project
- ☐ **PT3** Selecting Supply Chain Members at the Project Level
- ✔ **PT4** Implementing Integral Value Engineering on a Project

- ☐ **PO1** Applying ADePT to Design Management
- ☐ **PO2** Applying DePlan to Design Management
- ☐ **PO3** Modelling Project Design Processes
- ☐ **PO4** Assembling Supply Chains
- ✔ **PO5** Applying Value-adding Tools to Design Problems

Identifying and Reviewing Applications

The application of tools needs be to monitored to record how they have been used. Spontaneous feedback should be encouraged, but prompts are required to capture information. Where relationships between individual tools, or between tools and particular types of technical design problems emerge, these should also be investigated and recorded in the value-adding toolbox (the value-adding toolbox is made available within projects by the Applying Value-adding Tools to Design Problems (PO5) and Implementing Integral Value Engineering on a Project (PT4)).

Record the key characteristics of each application and how tools have been combined to solve a design problem. To be most effective, these examples also need to contain additional information describing the context and reasoning behind the choice of a particular combination.

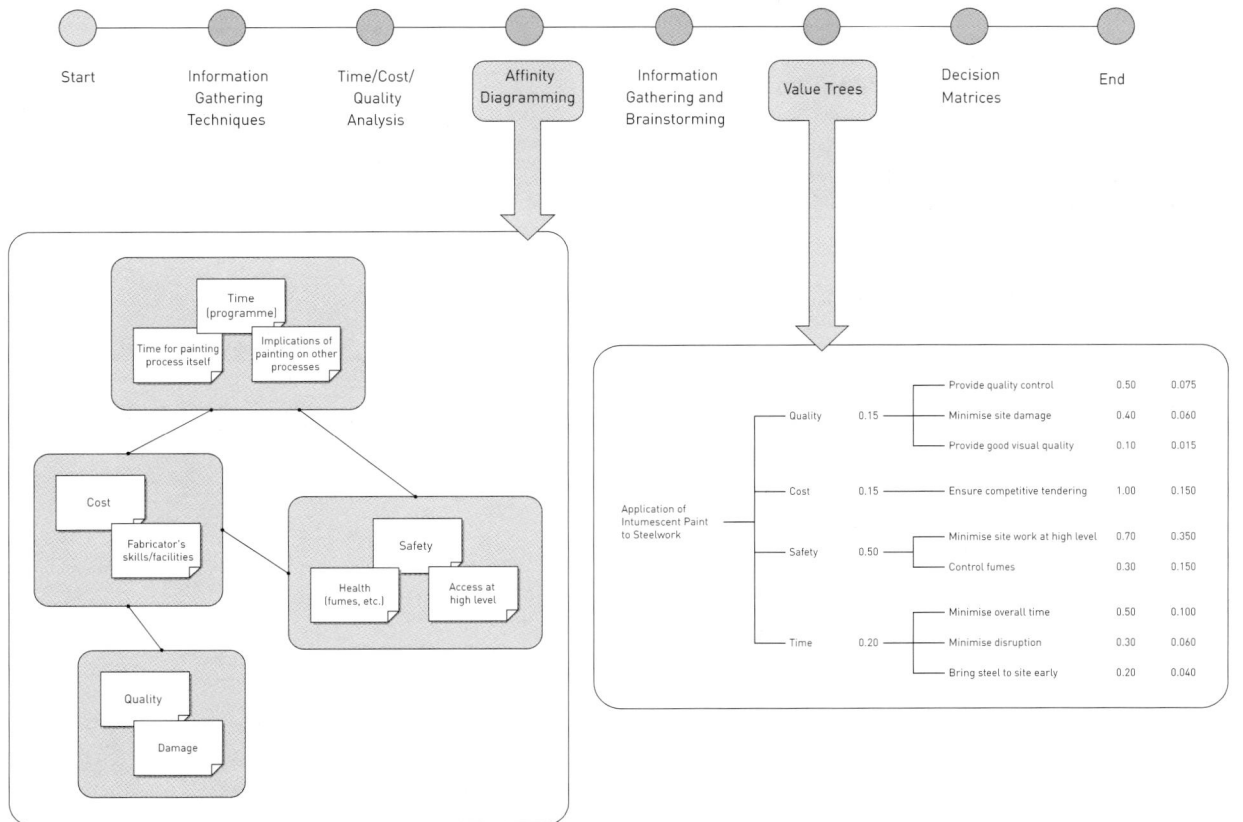

Capturing Informative Examples

Having reviewed the applications of tools on a project, appropriate aspects are documented to be fed back to subsequent projects and other value system members. The reporting of both good and bad (the label often being viewed as failures) requires careful management. To overcome the natural reluctance to share bad experiences, a learning culture must be established where all examples are viewed as informative lessons, rather than as critical analyses of what went wrong and what went right.

Case histories can be retained within a toolbox maintained by a single organisation, but if it is a collaborative resource, the knowledge will be shared among all organisations. This may be seen as a reward for collaboration in maintaining the toolbox, as it gives them access to broader knowledge than they could accumulate by themselves.

It is a good idea to index the information in a way that aids searching for suitable tools on subsequent design problems. This should take account of the five methods of tool selection (see the section on Selecting Suitable Tools in Applying Value-adding Tools to Design Problems (PO5)).

Disseminating Case Histories

Feedback on value-adding tool applications should be documented in a standard format for incorporation into the toolbox for dissemination. A standard format allows examples to be compared and helps in future tool selection. This documentation also constructs an audit trail of the contributions that design organisations make to the project, demonstrating to the client the value of the design.

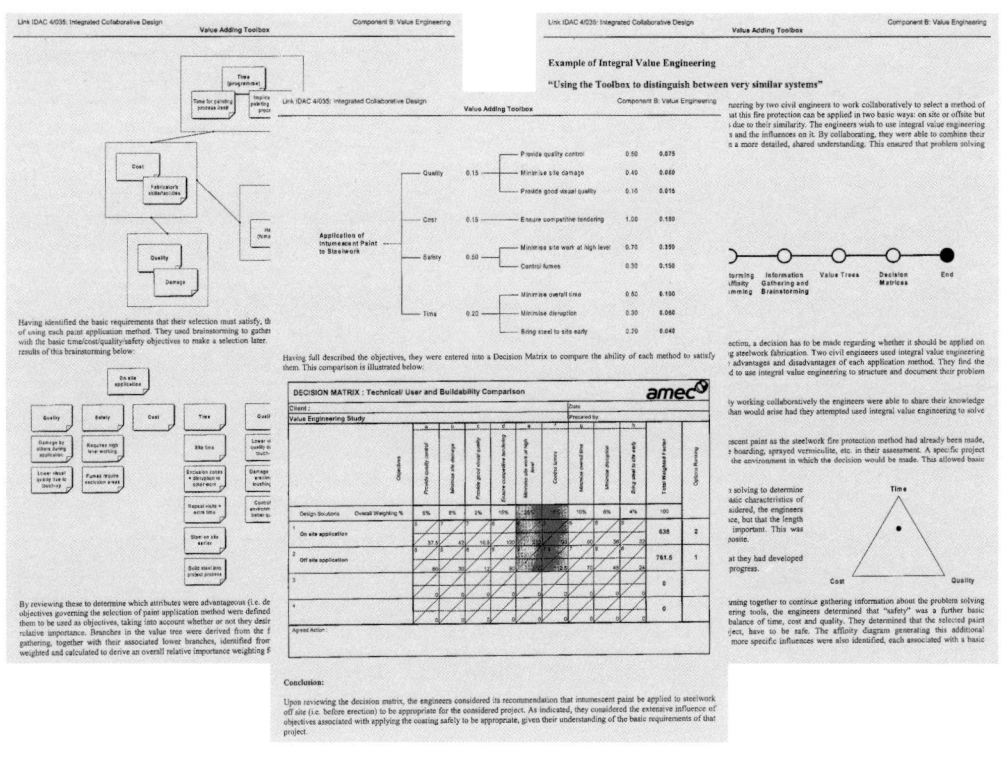

PS1

Planning Project Design Management

Provider benefits

- Linking project design management practices to the business strategy of all design chain members will result in project transparency, where providers will have a clear understanding of what is required by the receiver.

- Providers can also develop an improved understanding of their interfaces with other providers in a particular project design chain.

Purpose

This practice is used to plan how design management practices are deployed within the project domain, and to help design chain organisations operate in a manner that satisfies their business objectives. Organisations will be able to improve their understanding of project processes and subsequently improve the way in which processes can be represented in repeated applications of business process models to projects.

Summary

An organisation should have an understanding of its generic processes prior to becoming a member of a design chain (see Planning Design Process Management (BS1)). This will ensure that organisations are able to define their design information needs and communicate them to other design chain members (see Modelling Project Design Processes (PO3)). Also, if each organisation is aligned within the supply network, then design chain alignment will be easier to achieve. If the supply network has not matured, and alignment is not complete, then emerging design chain organisations will need to model their business processes (see Modelling Business Design Processes (BO1)). Design management links the business need of adopting a process view to understand how organisations are aligned to form design chains in the project domain, and it provides the management regime that supports the design chain operation. The planning of project design management practice is concerned with supporting the need to understand the business design process, ensuring that they are representative of the project concerned, and applying these models to design projects.

Receiver benefits

- Receivers will be able to clearly communicate their project needs for the alignment of cost planning, risk and design information to the providers.

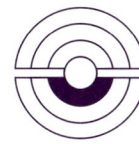

Outline Procedure

1. Where are we starting from?
2. What do we want to achieve?
3. How will we do it?
4. How well are we doing?

Requirements and Resources

- Although this practice is applied in the project domain, it is directly related to the business domain strategy. It should therefore be applied in conjunction with Planning Design Process Management (BS1).
- An individual from within each organisation in a supply chain should be given the responsibility for their design management activities, thus ensuring that the project strategy is developed in conjunction with the business needs.

Related Practices

- ☑ **BS1** Planning Design Process Management
- ☐ **BS2** Planning Supply Chain Management Business Practice
- ☐ **BS3** Planning the Implementation of Integral Value Engineering across the Business

- ☐ **BT1** Applying Process Management in the Business
- ☐ **BT2** Aligning Supply Networks
- ☐ **BT3** Applying Supply Chain Management in the Business
- ☐ **BT4** Applying Integral Value Engineering in the Business
- ☐ **BT5** Conducting a Value Survey
- ☐ **BT6** Performing an ADePT Review

- ☑ **BO1** Modelling Business Design Processes
- ☐ **BO2** Auditing the Supply Network
- ☐ **BO3** Auditing the Supply Network for Technical Competence
- ☐ **BO4** Gathering Value-adding Feedback from Projects

- ☐ **PS1** Planning Project Design Management
- ☐ **PS2** Planning Supply Chain Management Project Practice
- ☐ **PS3** Planning Integral Value Engineering Project Practice

- ☑ **PT1** Applying Design Management Practices
- ☐ **PT2** Applying Supply Chain Management to a Project
- ☐ **PT3** Selecting Supply Chain Members at the Project Level
- ☐ **PT4** Implementing Integral Value Engineering on a Project

- ☐ **PO1** Applying ADePT to Design Management
- ☐ **PO2** Applying DePlan to Design Management
- ☑ **PO3** Modelling Project Design Processes
- ☐ **PO4** Assembling Supply Chains
- ☐ **PO5** Applying Value-adding Tools to Design Problems

Where are we starting from?

This stage is concerned with gathering information on project operations and practices to inform your strategic planning for projects.

The practice of project design management across the design chain should be reviewed to determine how your organisation currently works with other organisations in the project domain. These relationships are likely to be mediated through some form of contract, but these need to be augmented with clear definitions of responsibilities and interfaces. Forming a real understanding of the information needs of design chain partners will be crucial in developing effective project design process management strategies. This will help you to develop strategies to build upon these existing ways of working to achieve greater collaboration and integration in the design process. The project team might consider undertaking a collaborative assessment of their ICD process management maturity using Table 3.2 in Section 3.5.

Working practices should be reviewed to identify how project design management is undertaken, including the various process management resources. Your project team will need to determine its understanding of:

- your design process management practices, defined from your business domain through formal procedures, job descriptions and staff development;

- project design processes, including any experience of adopting a process view as a business and as part of a design chain;

- the team's experience of operating as part of a design chain, and understanding the maturity of an ICD approach;

- project applications of a shared, high-level process framework that defines project stages and the main activities undertaken at each stage;

- project applications of low-level design process maps, showing how your organisation undertakes design activity in detail, including existing approaches to work-breakdown structures and project design programming;

- your current capability in design management - identifying the resources allocated to project design management and assigning roles and responsibilities to key individuals in the team;

- how you currently manage your design process, both in the project and business domains, with particular reference to external design interfaces; and

- the potential barriers to change and improvement of your project design management practice.

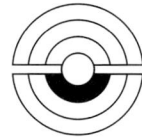

What do we want to achieve?

This stage is concerned with setting appropriate project goals, in light of the current status of design management within your organisation and in the supply chain.

ICD organisations will need to develop a project strategy to adapt and deploy a process framework that has been developed as part of their business operations (see Modelling Business Design Processes (BO1)). A process framework will have been deployed to suit their business processes, but it needs to be adapted to suit the characteristics of each project (see Applying Design Management Practices (PT1)). Also, a strategy for modelling project design processes will need to be planned to ensure that those ICD organisations that form part of a design chain understand the scope of design activity in a particular project. Thus, the design chain must align its project design processes with a clear demarcation of roles and responsibilities, and streamlining the flow of information.

The project strategy for integrating a supply chain around a common view of the design process will be dependent on the maturity of the ICD approaches within the project design chain. The overall objective will be to define processes and interfaces to assure a smooth and effective flow of timely information, with a well-defined set of design responsibilities. The timing and scale of design chain input to the project strategy is important to achieve a balance between appropriate influence on the processes aligned with selection and appointment activities in the project domain.

The implementation of project design management needs to be planned in conjunction with other strategic developments and take into account all parts of an organisation's design management approach (see Planning Design Process Management (BS1)). In the project domain, the design chain is formed by a number of supply network members, with varying levels of maturity of the ICD approach. Therefore, it is important for design chain organisations to adopt a coherent process view and agree a project design management approach that will suit the needs of the project (see Applying Design Management Practices (PT1)).

How will we do it?

This stage is concerned with planning a logical development of your working methods to implement your process management approach on projects.

Define Strategic Scope
Define the required scope of change by comparing the project team's current project design management practice with those required. This will determine the extent of the change in your project team's working practices that your project strategy must realise. The opportunity to influence the scope of design process management will obviously be reduced if your organisation has entered the project after other design organisations.

Select and Plan Deployment Method
Process management is deployed through a combination of leadership by design managers and the communication of process descriptions, responsibilities and plans. The design teams should be briefed at the earliest opportunity on the process framework for the project and their responsibilities within it. It is important to engage as much of the design chain as possible, perhaps through a workshop (or series of workshops for large teams). This will present an opportunity to test alignment of processes and to achieve a consensus in understanding the language and definitions to be used for the project stages and stage-gates. If the design chain is relatively immature in terms of ICD practice then it may be beneficial to ensure that the issues and resources required to establish the design management strategy are identified during the design chain selection process.

Graphical and textual process descriptions (maps) can be circulated in paper form, but electronic transmission via intranets, the web or a project extranet is more efficient and is increasingly becoming the norm. Some tools, such as the Process Protocol (see Figure 2.8) and modelling software, directly support such an approach.

Team meetings, such as design reviews, should be driven, scheduled and informed by the design chain's definition of the design process. This activity should be supported by shared procedures, combined where appropriate with those specific to individual organisations.

Each organisation will need to ensure that they have adequate resources available to adopt and deploy the process framework and models, and individuals will need training.

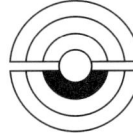

How well are we doing?

This stage is concerned with monitoring the effect your strategy is having, so that you learn from its deployment, implement corrective action within the project, and improve your project operations through feedback to your business operations.

Determine Feedback Required
Your strategy should be adaptable during deployment to allow for any obstacles encountered. Identify the key performance measures, especially those associated with the strategic targets for the project.

Establish Feedback Mechanisms
The repeated application of a process framework and project design process models to projects will inevitably yield improvements to the generic processes developed in the business domain (see Planning Design Process Management (BS1)). These improvement will need to be fed back to the business domain, where they can be incorporated to ensure that the benefits are captured and passed on to the next application to the project domain. Each organisation will need to plan the necessary mechanisms to manage the feedback process, and provide the necessary resources to ensure that it is undertaken in the appropriate manner.

Gather and Respond to Feedback
Once project design management practice is established and underway, it is important that it remains appropriate to the operating project environment. In a similar manner to responding to change when establishing design management practice, this continuous review helps the organisation determine if its emerging design management practice remains compatible with the project objectives and with the latest operating needs of the business.

PS2

Provider benefits

- Ensures that the commercial and management frameworks within which providers have to operate are appropriate to the project and to the provider's involvement in it.

Receiver benefits

- Allows the application of SCM to be flexible enough to allow generic processes to be adjusted to suit a project's individual characteristics and criteria.

- Provides an opportunity for a receiver to review the means of incentivising the providers within the design chain and incorporate these in the project's management and commercial framework.

Purpose

To ensure that the project undertaken in the project domain is planned in line with the agreed SCM strategy in the business domain. This practice is particularly concerned with how design chains are assembled from the supply networks. While this may be of more practical concern to receivers than providers, both roles have a responsibility in design chains to ensure strategy at the business level is implemented on individual projects.

Summary

This practice helps organisations plan projects strategically. It enables managers to structure their project implementation plans along ICD principles and to build design chains within them. The techniques for doing this include:

- project partnering - creating a project-specific partnering framework for providers and a receiver;
- creating the commercial structure for a project that will best facilitate the alignment of the providers' and receivers' commercial interests with each other and with the objectives for the project;
- tendering strategies - selecting the appropriate procurement path and tendering strategy for each work package; and
- tailoring generic processes to project needs and ensuring that the generic project processes are modified to suit specified project criteria.

As with other practices, the list is not exhaustive, but illustrates the types of issue that must be taken into account.

Outline Procedure

Because many of the techniques in this practice are interrelated, such as project partnering and the formation of an appropriate commercial structure for the project, it is helpful to approach the development of a strategy in four stages:

1. Where are we starting from?
2. What do we want to achieve?
3. How will we do it?
4. How well have we performed?

Requirements and Resources

- No additional resources are required for this practice as the issues and techniques addressed represent good practice in all projects. However, additional information on the tactical application of these techniques can be found in Applying Supply Chain Management in the Business (BT3).

Related Practices

- ☐ **BS1** Planning Design Process Management
- ☑ **BS2** Planning Supply Chain Management Business Practice
- ☐ **BS3** Planning the Implementation of Integral Value Engineering across the Business

- ☐ **BT1** Applying Process Management in the Business
- ☐ **BT2** Aligning Supply Networks
- ☑ **BT3** Applying Supply Chain Management in the Business
- ☐ **BT4** Applying Integral Value Engineering in the Business
- ☐ **BT5** Conducting a Value Survey
- ☐ **BT6** Performing an ADePT Review

- ☐ **BO1** Modelling Business Design Processes
- ☐ **BO2** Auditing the Supply Network
- ☐ **BO3** Auditing the Supply Network for Technical Competence
- ☐ **BO4** Gathering Value-adding Feedback from Projects

- ☑ **PS1** Planning Project Design Management
- ☐ **PS2** Planning Supply Chain Management Project Practice
- ☑ **PS3** Planning Integral Value Engineering Project Practice

- ☐ **PT1** Applying Design Management Practices
- ☐ **PT2** Applying Supply Chain Management to a Project
- ☐ **PT3** Selecting Supply Chain Members at the Project Level
- ☐ **PT4** Implementing Integral Value Engineering on a Project

- ☐ **PO1** Applying ADePT to Design Management
- ☐ **PO2** Applying DePlan to Design Management
- ☐ **PO3** Modelling Project Design Processes
- ☑ **PO4** Assembling Supply Chains
- ☐ **PO5** Applying Value-adding Tools to Design Problems

Where we are starting from?

This stage is concerned with gathering information to inform your strategic plans for the project.

Review Assessment of ICD Principles in your Organisation

Your project strategy should be developed with regard to the current status of ICD principles within the project team. It may be beneficial to conduct a joint review of the level of maturity among the project team using the assessment from Table 3.3. This should indicate the project organisation's level of maturity, and act as the context in which individual tactics on projects have to be delivered.

What is occurring in the business domain?

In addressing the starting point for the strategy for implementing SCM at a project level, the strategic managers of the project need to take account of what is occurring in the business domain that can be harnessed for the project's benefit.

Project Partnering

The pre-existing relationships that may exist will influence how organisations chose to engage with one another. The existence of supply networks both formal and informal will constrain the approaches available to project personnel.

Tendering Strategies

Organisations may have preferences for working in particular ways with preferred tendering strategies or policies governing their engagement with other organisations to deliver projects, and this is a factor in the setting of any project strategy.

Generic Processes

The existence of generic processes developed in the business domain (such as an organisation's design process or formalised explicit ways of working) will form a pre-existing framework and resource for the development of a project strategy for the adoption of SCM practices.

How has the project evolved to date?

While projects may often have formal starting points, very rarely does this represent the first point of contact between organisations who may come together to form a design chain. Informal discussion exist, clients and end-users express preferences, and organisations who have specific competencies may be providing informal input to answer technical queries regarding the availability of materials and lead times without specific knowledge of a particular project. All these discussions may occur prior to the commencement of a formal project delivery process, but will form a backdrop to the development of strategies for delivering that project.

What do we want to achieve?

This stage is concerned with setting appropriate project goals, in light of the current status of adopting SCM practices within your organisation, and the time period for development. At the strategic level of the project, the intention should be to develop a project strategy or approach to the project that will:

- present a rationale for the selection of procurement strategies;

- provide support for the definition and allocation of work packages (drawing on generic work packages identified through Applying Supply Chain Management in the Business (BT3));

- set limits to the autonomy of individuals or organisations to act - in part, setting the limits to the strategic plan may be decided at the business level, but individual project managers and design managers may be authorised to extend these;

- establish the core project procedures (which would typically be mandatory) and determine the guidelines (which might be discretionary) for how the project should be managed; and

- communicate the management philosophy to the project team - even though organisations (and often individuals) might be selected on the basis of exhibiting a similar management philosophy, there is always a need to express that philosophy and provide a means of communicating it to others.

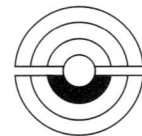

This stage is concerned with planning the development of your design chain to deliver the project in accordance with the ICD principles. Four main elements need to be addressed in the development of the project strategy for implementing SCM practices:

Project Partnering

Project partnering, in its broadest sense, involves an agreement by two or more parties to work collaboratively on a project[1] . Partnering can take many forms, such as joint ventures or consortia arrangements. Where ICD differs is that while project relationships are defined by contractual conditions, relationships are also influenced by trust, mutual respect, and the shared understanding of each partner's expectations and values. This can only happen when the cultures of the collaborating organisations are aligned with each other and the importance ascribed to organisational boundaries adjusted to make room for a collaborative culture and an alignment of objectives.

With ICD, partnering at the project level should build upon the collaborative effort that occurs at the business level. ICD seeks to make these business arrangements (which are often informal) explicit and either relate them to the characteristics of particular projects, or develop additional arrangements to take into account a specific circumstance (e.g. a joint-venture arrangement). In addition, project partnering can be used as a mechanism to draw providers, who are outside of the supply network, into the culture of the network for the duration of the project.

Creating an Appropriate Project Commercial Structure

This deals with the translation of the strategy developed in the business domain into the appropriate design of the project arrangements between organisations and ensuring that it is appropriate for the project's environment. Issues that may be given consideration include the following:

- *Choice of procurement path*. Particular procurement paths can create barriers to effective design chain integration by creating disincentives to working in a collaborative fashion and even creating incentives to act against the ICD ethos. The practices incorporated within the ICD approach are capable of being applied to any procurement path but the procurement path will affect their effectiveness.

- *Use of open book*. Open book approaches are becoming more common in the construction industry as a proof of honesty in situations where trust in not normally present. In the early stages of supply networks, open book has a role in building trust but the need for it to be used in the project domain may decline over time as strategic procurement matures and moves towards order-driven (as opposed to contract driven) procurement.

- *Appropriateness of contractual arrangements*. The importance of design contribution undermines a 'one-size-fits-all' policy towards contractual arrangements. Instead any project will utilise a range of procurement paths and contractual arrangements. Key strategic work packages are likely to be procured differently to other packages where design input from the provider is less critical. In addition, incentivisation arrangements are necessary to ensure rewards for all parties in achieving overall project objectives.

Tailoring Tendering Strategies to Projects

An established supply network creates a business environment that allows considerations other than the lowest bid when making tenders. Issues, such as reliability of delivery, quality or compatibility with other organisations, may also be important criteria reflecting the project objectives, as well as procurement strategies and the financial structure of the project.

Within a supply network, techniques used to appoint supply chains vary depending on the nature and the value of the product. High-value and high-risk provider input is less suited to competitive tendering, whereas standard products may well benefit from such an approach. The four-box matrix used in Planning Supply Chain Management Business Practice (BS2) can be used to identify a suitable tendering strategy for each work package.

Continued on next page

[1]Bennett, J., Hayes, S. (1998) *The Seven Pillars of Partnering: A guide to second generation partnering*, Thomas Telford, London.

Continued from previous page

An objective of SCM is to create fewer and deeper relationships between providers and receivers. As a consequence, providers expect an increased workload from the receivers with whom they collaborate to justify the investments needed to work in a supply network. The enhanced certainty of workload and the reduction in costs (e.g. tendering) should free up resources for more value-adding activities.

Tailoring Generic Processes to Project Requirements

Few generic processes are directly applicable to projects without tailoring to meet specific needs. Examples of this are given in Planning Project Design Management (PS1) and Planning Integral Value Engineering Project Practice (PS3).

In the early stages of a project, the receiver will have to review the generic processes to see which can be modified or used for the project in question. This review process is described in Assembling Supply Chains (PO4). Although this activity may be seen as one specific to the receiver, all organisations will need to assess how their internal project processes will need to be adapted so that they can integrate effectively with each other and contribute to the project.

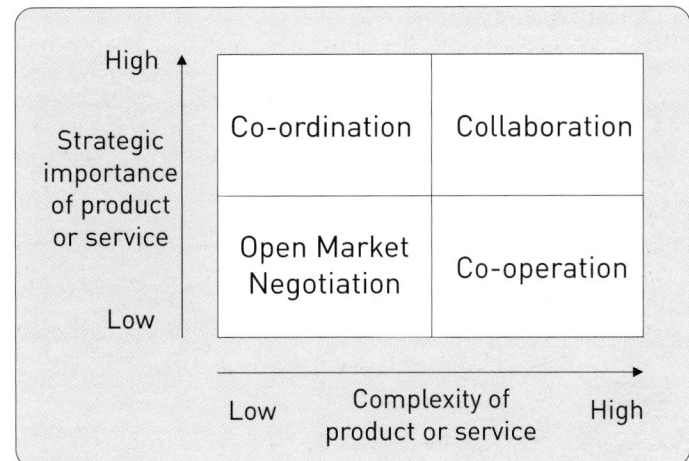

Providers

- With ICD, providers should be able to influence the types of procurement path being considered by the receiver, building upon relationships in the business domain.

- By receivers concentrating on fewer potential providers, the providers' overheads for unsuccessful tenders should reduce. Resources can therefore be allocated to areas such as cost planning in the early stages of individual projects resulting in greater predictability of costs to clients.

Receivers

- ICD creates an opportunity to select procurement paths in a way that allocates or shares risk on an equitable basis in pursuit of common project objectives. The design chain can then concentrate on reducing the likelihood of an undesired outcome to the project occurring rather than concentrating on who will be held responsible for the consequences of that outcome should it happen.

- Just as risks and rewards may need to be shared or allocated between providers and receivers, so too must costs and benefits be spread throughout the design chain in a transparent fashion. This enables providers to make the appropriate decisions and modifications to their behaviour to ensure that working in a collaborative fashion is in their interests.

- Reduced tender lists reduce the cost of tendering for providers and simplify the adjudication process for the receiver. In addition, through the use of supply networks, the selection of providers can be targeted more effectively because the receiver will have a better understanding of the provider's capabilities.

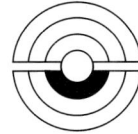

How well have we performed?

This stage is concerned with monitoring the effect your strategy is having, so that you can take corrective action, if required. The primary means to achieve this is through project reviews that can be held at stages through a project's life (e.g. at stage gates) or on completion. The type of review that will need to be undertaken is described in Assembling Supply Chains (PO4).

- *Project Partnering*. The aims and intentions of project partnering are compatible with supply chain management theory. In practice without the supporting business relationships, there is a tendency for the high standards of project partnering to slip into a hybrid between a partnering approach and traditional working practices where the partnering ethos can be abused, particularly when performance difficulties arise. Monitoring of the project partnership through the project review process will provide an opportunity to assess whether this is occurring.

- *Project Commercial Structure*. The appropriateness of the project's commercial structure is less amenable to change as the project process develops. However, project reviews can provide lessons as to what has worked and what has not.

- *Tendering Strategies*. Project reviews can indicate where strategies have gone wrong. Early indications can be identified in the project and since tendering occurs over a prolonged period of time in a project's life cycle, steps can be taken to avoid mistakes that have been made to gain the benefit achieved where work packages or elements of design work have already been let.

- *Tailoring Generic Processes*. The review and modification of project process should be an ongoing activity. The failure to modify processes in the light of experience on a project will result in the undermining of the legitimacy of the processes, such that procedures are ignored and ad hoc working methods develop.

PS3

Planning Integral Value Engineering Project Practice

Provider benefits

- Linking IVE project practice to the business strategies of other design chain members gives other providers access to the full breadth of IVE experience embodied in value-adding toolboxes.

Purpose

The planning of Integral Value Engineering (IVE) in the project domain must be consistent with the strategic planning that has taken place in the business domain. Organisations with project management responsibility use this practice to make IVE resources available to all organisations in a design chain. This included planning how a value-adding toolbox (assembled in the business domain) can be used on projects.

Summary

The planning of IVE implementation across projects is mainly concerned with the use of value-adding toolboxes on projects and making sure their content is relevant. This is achieved by maintaining the collections of value-adding tools and examples of their application. If members of a value system collaboratively maintain a single toolbox, then it becomes a common resource for project design chains involving those organisations. Implementing Integral Value Engineering on a Project (PT4) gives a number of methods of introducing IVE resources to the project domain. IVE will be most effective when the project values (e.g. objectives or functional requirements) are communicated throughout the project team. Typically, these will build from the use of value management during the inception and conception of projects to include methods of maintaining awareness of them throughout project progression.

Receiver benefits

- IVE practices help receivers manage project activity in line with their long-term business strategies. With ICD, the practices of Planning the Implementation of Integral Value Engineering across the Business (BS3) and Applying Integral Value Engineering in the Business (BT4) can both help with the development of such strategies, particularly in how they can support providers in a supply network or a value system.

This practice outlines three types of toolbox - shared, rented and owned - and the planning of their deployment, application and capture of learning. This strategy and its related tactic (PT4) are mainly concerned with the first two types, extending the business domain practices (BS3 and BT4) to groups of organisations.

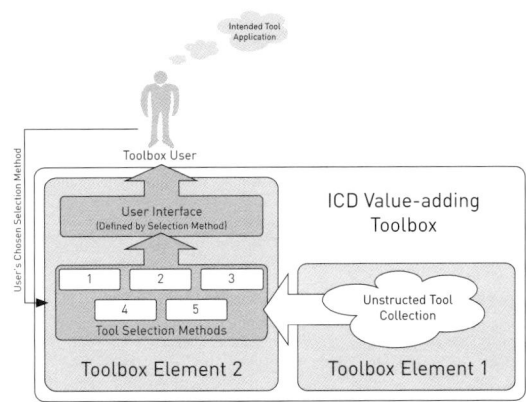

Project Deployment of an IVE Resource

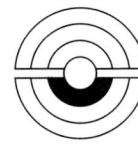

Outline Procedure

1. Where are we starting from?
2. What do we want to achieve?
3. How will we do it?
4. How well are we doing?

Requirements and Resources

- IVE must be used in conjunction with several business domain practices. One such business domain practice is the development of value-adding toolboxes, which requires maintenance of a database of value-adding tools and lessons learned as the basis for continual improvement.

Related Practices

- ☐ **BS1** Planning Design Process Management
- ☐ **BS2** Planning Supply Chain Management Business Practice
- ✔ **BS3** Planning the Implementation of Integral Value Engineering across the Business

- ☐ **BT1** Applying Process Management in the Business
- ☐ **BT2** Aligning Supply Networks
- ☐ **BT3** Applying Supply Chain Management in the Business
- ✔ **BT4** Applying Integral Value Engineering in the Business
- ☐ **BT5** Conducting a Value Survey
- ☐ **BT6** Performing an ADePT Review

- ☐ **BO1** Modelling Business Design Processes
- ☐ **BO2** Auditing the Supply Network
- ☐ **BO3** Auditing the Supply Network for Technical Competence
- ✔ **BO4** Gathering Value-adding Feedback from Projects

- ☐ **PS1** Planning Project Design Management
- ☐ **PS2** Planning Supply Chain Management Project Practice
- ☐ **PS3** Planning Integral Value Engineering Project Practice

- ☐ **PT1** Applying Design Management Practices
- ☐ **PT2** Applying Supply Chain Management to a Project
- ☐ **PT3** Selecting Supply Chain Members at the Project Level
- ✔ **PT4** Implementing Integral Value Engineering on a Project

- ☐ **PO1** Applying ADePT to Design Management
- ☐ **PO2** Applying DePlan to Design Management
- ☐ **PO3** Modelling Project Design Processes
- ☐ **PO4** Assembling Supply Chains
- ☐ **PO5** Applying Value-adding Tools to Design Problems

Where are we starting from?

The first stage of this project strategy must determine which IVE resources are available within the project environment, and whether they are capable of supporting IVE. This provides the background information dictating the scope of IVE practice and the necessary resources. The assessment in Table 3.4 of Section 3.5 can be used to structure a project team's assessment of the status of current IVE practice in each of their organisations.

If the project design chain is assembled from organisations who are (mostly) members of a common value framework, then the need for this review will be reduced, as framework members will have prior knowledge of each other's IVE resources before the project commences. Ideally, the value frameworks will share and maintain a common set of IVE resources in the business domain that can be directly applied to projects as and when they arise. If this is the case, then the purpose of the strategy is to adopt those resources and plan their project use to meet the specific requirements of the project considered. If this is not the case, then this strategy must also identify and assemble IVE resources before planning their deployment on the project.

What do we want to achieve?

Identify Required Project Resources

The nature of the project should be reviewed to identify the most appropriate approach to IVE practice within it. For example, does the project contain any system or elements that are particularly complex or specialised? Will these require specific attention during their delivery? You should then identify the required IVE resources and check if they are currently available from the project partners.

Seek Additional Resources

If some IVE resources are not available among the project partners, they may be obtained by the following:

- Inviting organisations possessing the required resources into the current project. If the resource is essential, then the organisation may also be invited into the supply network or value framework in the business domain (provided what other criteria for membership are satisfied). This approach requires careful management and design responsibility may have to be reallocated to integrate them into the design chain.

- Developing the required resources within the current project, for sharing by all project partners. Make sure you build and maintained these resources so that they can be fed back to the business domain for application on subsequent projects.

- Procuring the additional resources from a third party for the duration of the project. This approach does not facilitate feedback and learning.

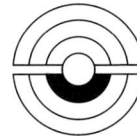

How will we do it?

Define Strategic Scope

Two approaches, whose suitability depends on the nature of the organisation and project are:

1. allowing resources to be used in a spontaneous, unstructured way that responds to project circumstances. With this approach, it is important to establish a common understanding of how the resources can be used to help with the design. One way of achieving this is to give basic training in the value-adding toolbox to the designers in all the companies when they join the project; and

2. establishing IVE practice in line with an organisation's business strategy and then manage the resource use accordingly. One way of doing this is to regularly review the problem-solving tools in the toolbox to ensure that the tools being developed are taking the company and the project in the right direction.

Select Deployment Method

IVE resources will be most effective if they are made available to all project members, at all levels of their organisation. This is particularly important in the case of project designers, who must have ready access to the resources to apply them to their everyday work. With the value-adding toolbox the actual resource is information and this is commonly exchanged electronically, by company or by project intranet (see figures below).

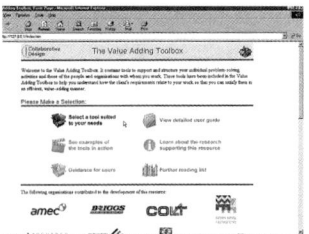 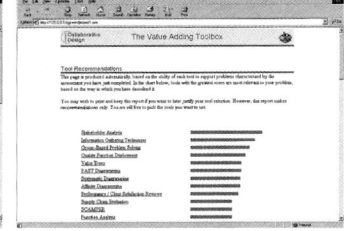 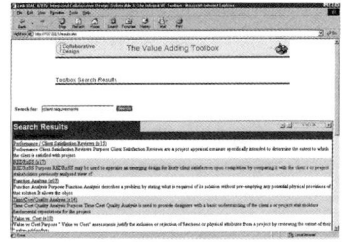

Plan Deployment

The strategy must define the methods of introducing appropriate IVE resources from the business domains of project members. Planning deployment at the outset of a project allows the project partners and environment to set fully and appropriate resourcing objectives. Project tactics are also developed at this time to define how these resources will be made available for use within each project (see Implementing Integral Value Engineering on a Project (PT4)). The plans will normally focus on the value-adding toolbox and the collection of feedback (see Gathering Value-adding Feedback from Projects (B04)).

A toolbox can be used in three ways: shared, owned or rented. The first is where a single toolbox is shared within a project design chain and, by means of appropriate project IT, made available to all members. The second is when an organisation deploys a toolbox for its own use over its intranet. Finally, a toolbox can be made available to projects via a third party. However, in this case, the toolbox content is generic and may be less useful than a shared or owned version.

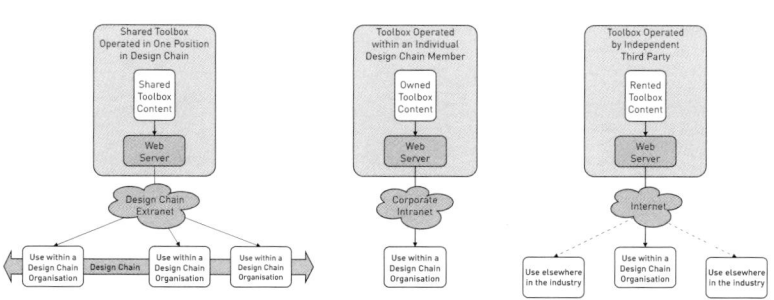

Receivers

- If a receiver is planning to share IVE resources with all project members, it will need to investigate whether the resources are compatible with the systems being used by the project members and whether they are in a format that can be incorporated into project working practices.

How well are we doing?

Determine Feedback Required
Strategies must make provisions that will allow the lessons and experience from projects to be identified, gathered and fed back to the business domain. By doing so, this information can become corporate knowledge, improving the effectiveness with which IVE resource can be deployed subsequently. See Gathering Value-adding Feedback from Projects (B04) for a more detailed explanation of this requirement. In the case of a value-adding toolbox, for example, the feedback should provide sufficient information to maintain and enhance the content.

Establish Feedback Mechanisms
Where appropriate, these should adopt the same mechanism/medium used for deployment. The key is to make it quick, easy and structured. There may need to be a balance struck in the quantity of information sought. You must also decide who will have responsibility for reviewing and responding, and aim to make their job straightforward.

Gather and Respond to Feedback
The information being captured should be reviewed regularly. Users need to be encouraged that their feedback is being used constructively; this can be achieved by updating the content of the resources (perhaps with a notification) and by personal responses to individuals. It will also serve the project's cause if positive feedback and examples of adding value are communicated to the client and stakeholders.

If the information does not sufficiently inform monitoring of this project strategy, then action will be required to revise how it is being captured (if this is preventing appropriate feedback being gathered) or to provide further training and encouragement to users.

Providers

- When a provider is partnered with other organisations, the lessons learned should be shared to take everyone forward from a common platform. Because feedback can be negative as well as positive, care must be taken to reassure any provider delivering bad news that to so will not endanger its position in the value system. If providers feel threatened in reporting tools that have not worked as expected, the whole system/design chain will miss out on vital information that could affect the outcome of future projects.

Receivers

- Corporate learning can profit from project innovation that is passed up the chain. All can benefit if this is shared with other value system members.

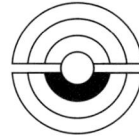

PT1

Applying Design Management Practices

Provider benefits

- By adopting a process view of their project operations, providers will be able to define their high-level and detailed level processes, creating the opportunity to develop a common framework of language that can be used by the whole organisation, and between other supply network members.

Receiver benefits

- Receivers can consolidate their understanding of their organisational business and project processes, thus streamlining internal processes and improving alignment with other members of the project design chain.

- Receivers will be able to communicate their project information needs to other supply network members, therefore improving their external alignment.

Purpose

To provide support for the deployment of design management practice within the design chain, ensuring that each organisation has the necessary guidance to be able to create and maintain an ICD environment. This will provide individual designers with the opportunity to plan and manage design based upon their detailed understanding of their processes and how their outputs are used by other members of the design chain.

Summary

Supply network organisations may have different levels of understanding of their individual processes within the business domain (dependent upon the maturity of the supply network) and, therefore, will need to determine how to agree on a common project language and approach to design management. The project organisation should collectively determine the requirements of the project, as this will influence the design management approach for the design chain, and will lead to a clear understanding of the demarcation of design activities. ICD organisations need to review the available design management resources prior to planning their operational roles within a design chain. Provisions need to be made to monitor the resources used when managing the design process, thus ensuring that any process management learning is fed back to the business domain, where improvements can be incorporated into the generic business process models. ICD organisations will then be ready to plan the use of project operations to assist them in the planning and management of the design process.

Outline Procedure

1. Determine Results Required
2. Plan and Develop Approaches
3. Deploy Approaches
4. Assess and Review Approaches

Requirements and Resources

- Organisations should have adopted the ICD principle Adopting Process Management and the practices of Planning Design Process Management (BS1), and Planning Project Design Management (PS1).
- Suitable mechanisms are required to communicate the need to adopt a process view of projects, including start-up workshops, project handbooks and electronic media.
- Each organisation should appoint an individual who is responsible for managing the deployment of resources and co-ordinate the planned deployment of project operations.

Related Practices

- ✔ **BS1** Planning Design Process Management
- ☐ **BS2** Planning Supply Chain Management Business Practice
- ☐ **BS3** Planning the Implementation of Integral Value Engineering across the Business

- ✔ **BT1** Applying Process Management in the Business
- ☐ **BT2** Aligning Supply Networks
- ☐ **BT3** Applying Supply Chain Management in the Business
- ☐ **BT4** Applying Integral Value Engineering in the Business
- ☐ **BT5** Conducting a Value Survey
- ☐ **BT6** Performing an ADePT Review

- ☐ **BO1** Modelling Business Design Processes
- ☐ **BO2** Auditing the Supply Network
- ☐ **BO3** Auditing the Supply Network for Technical Competence
- ☐ **BO4** Gathering Value-adding Feedback from Projects

- ✔ **PS1** Planning Project Design Management
- ☐ **PS2** Planning Supply Chain Management Project Practice
- ☐ **PS3** Planning Integral Value Engineering Project Practice

- ☐ **PT1** Applying Design Management Practices
- ☐ **PT2** Applying Supply Chain Management to a Project
- ☐ **PT3** Selecting Supply Chain Members at the Project Level
- ☐ **PT4** Implementing Integral Value Engineering on a Project

- ✔ **PO1** Applying ADePT to Design Management
- ✔ **PO2** Applying DePlan to Design Management
- ✔ **PO3** Modelling Project Design Processes
- ☐ **PO4** Assembling Supply Chains
- ☐ **PO5** Applying Value-adding Tools to Design Problems

Determine Results Required

The requirements of the project will influence the design management approach taken by supply network members when forming a design chain, particularly the scope of the design activity and allocation of responsibility within the design chain. The allocation of design responsibilities should reflect the competencies of the design chain and the availability of key resources. This will help align their process view (adopted within their businesses) to the specific project domain, prior to the integrated planning and subsequent management of the design activities.

Design chain members will need to identify their training needs when adapting and applying a process framework and undertaking detailed level process modelling, so that individuals can be prepared to apply process management to the project domain.

Providers

- A clear understanding of the project design process and the division of responsibilities between the design chain members will help reduce abortive effort by providers.

Receivers

- Early communication of the design process to be adopted for a project will enable providers to determine their most effective contribution to the project.

Plan and Develop Approaches

The tactic of applying design management in projects is to engender a consistent process view in the project team and then deploy the required design management operations to assist in the planning and management of the design process. It is necessary for the emerging design chain organisations to review their existing design processes and to align themselves using a process framework and detailed design process models. A project workshop can help align the levels of understanding of the design process and the availability of organisational process models. The workshop can explore and fully understand the project requirements and select suitable models for the project including the stage-gates to be used. Gaps in the overall design process model can be plugged using the practice Modelling Project Design Processes (PO3).

Providers

- Alignment of expectations regarding deliverables at key project stage-gates will reduce overlaps and gaps between design chain members and should reduce abortive effort.

Receivers

- Clear definition of the tactics to be adopted for applying process management to a project should be based on an understanding of the capabilities and processes used by members of the design chain, supported by facilitated workshops and/or training in process management.

Deploy Approaches

Once the emerging design chain has determined the process model representing a particular project, the design chain should communicate project processes to all project team members. It is important that project processes are communicated to new ICD organisations that may join the design chain throughout the duration of the project. This can be achieved through the use of project posters, paper documents and project extranets.

Design chain organisations should use the project process models to agree the division of design responsibilities. Design planning and management practices, such as the production of design programmes and the subsequent management of design decisions and resources to ensure timely completion, should be based around the agreed processes and stage-gates. Organisations can use the project design process models (see Modelling Project Design Processes (PO3)) in conjunction with ADePT to plan their project design activities (see Applying ADePT to Design Management (PO1)), and the derived project programmes should be made available to all project team members. They should also consider the use of DePlan to enable design activities to be scheduled on the basis of availability of design deliverables and information flows (see Applying DePlan to Design Management (PO2)).

Assess and Review Approaches

Mechanisms should be established to foster project learning by feeding back process improvements to the business domain that can be incorporated into the business domain tactics and operations (see Planning Design Process Management (BS1), and Applying Process Management in the Business (BT1)). Project issues should be studied to determine whether process management has been misaligned, design responsibilities have been adequately defined, or project stage-gates are ill-defined. The design chain should ensure that all project-based learning is captured and exploited through the business domain practices. By doing this, ICD organisations will improve their effectiveness within the design chain environment (with future applications of the business models to projects), and maintain their organisational knowledge-based relationships within the supply network.

Providers

- The effectiveness of project design management tactics should be acknowledged and incorporated into the business domain.

Receivers

- Receivers should seek to build design chain relationships upon business relationships by effectively translating positive learning from projects into business strategies and tactics.

PT2

Applying Supply Chain Management to a Project

Provider benefits

- Providers benefit from improved work package definitions because these eliminate unnecessary work, improve communications in the design chain and deliver more efficient design.

- Providers also benefit from reduced contractual conflict in the supply chain.

Purpose

The practice of applying supply chain management (SCM) to projects is used to plan how design chains are assembled from an existing supply network and, then, how the chain is used to deliver a project.

Summary

This practice describes a number of issues that must be addressed by providers and receivers when they function as a design chain. It needs to be understood in conjunction with Planning Supply Chain Management Project Practice (PS2) and Assembling Supply Chains (PO4). Providers and receivers need to:

- understand how work packages are assembled to ensure design and management activities are allocated effectively;

- understand the common project culture and to managing expectations positively;

- establish mechanisms for communicating within the project and the design chain; and

- formally manage team-building activities as a way of improving the collaborative abilities of the design chain.

Many of these issues and mechanisms are not unique to SCM and are important issues in all projects. Within an ICD approach they assist the smooth functioning of the design chain.

Receiver benefits

- Receivers similarly benefit from the efficiencies that follow from improved work package definitions.

- SCM applied to projects means clear roles and responsibilities and rules of working – which all go towards reducing the possibilities for confusion and ambiguity.

- SCM is also a means of communicating desired behaviour (project culture) to all members of the design chain.

Outline Procedure

There are four stages in applying SCM to projects, but as one stage does not necessarily build on another, the order in which the stages are completed is not critical (although it is sensible to do some things before others). The order given below is the one that seems the most logical:

1. Defining and managing work packages
2. Establishing project ground rules
3. Selecting communication mechanisms
4. Using team-building techniques

Requirements and Resources

- It is important to ensure that SCM principles can be applied to projects consistently and without hindering other, established activities of organisations. This requires a standard method of applying SCM to projects, which may be developed individually by a single organisation or collaboratively by supply network members. See Planning Supply Chain Management Project Practice (PS2).

- Supply network membership must be established prior to the use of this practice (see Planning Supply Chain Management Business Practice (BS2) and Auditing the Supply Network (BO2)).

Related Practices

- ☐ **BS1** Planning Design Process Management
- ☑ **BS2** Planning Supply Chain Management Business Practice
- ☐ **BS3** Planning the Implementation of Integral Value Engineering across the Business

- ☐ **BT1** Applying Process Management in the Business
- ☐ **BT2** Aligning Supply Networks
- ☑ **BT3** Applying Supply Chain Management in the Business
- ☐ **BT4** Applying Integral Value Engineering in the Business
- ☐ **BT5** Conducting a Value Survey
- ☐ **BT6** Performing an ADePT Review

- ☑ **BO1** Modelling Business Design Processes
- ☑ **BO2** Auditing the Supply Network
- ☐ **BO3** Auditing the Supply Network for Technical Competence
- ☐ **BO4** Gathering Value-adding Feedback from Projects

- ☐ **PS1** Planning Project Design Management
- ☑ **PS2** Planning Supply Chain Management Project Practice
- ☐ **PS3** Planning Integral Value Engineering Project Practice

- ☐ **PT1** Applying Design Management Practices
- ☐ **PT2** Applying Supply Chain Management to a Project
- ☐ **PT3** Selecting Supply Chain Members at the Project Level
- ☑ **PT4** Implementing Integral Value Engineering on a Projects

- ☑ **PO1** Applying ADePT to Design Management
- ☐ **PO2** Applying DePlan to Design Management
- ☐ **PO3** Modelling Project Design Processes
- ☑ **PO4** Assembling Supply Chains
- ☐ **PO5** Applying Value-adding Tools to Design Problems

Defining and Managing Work Packages

The way work packages are defined can be used as a basis for allocating design and for ensuring provider and receiver activity is aligned so the project progresses efficiently. In traditional tendering, for example, work package definitions reflect what the receiver understands to be the most efficient way of grouping design tasks together. With this approach, providers have to compete with each other in demonstrating that they have the necessary technical and managerial skills. In adopting an ICD approach, work packages can be defined in ways that reflect the different combinations of organisations and allocation of design tasks between them, and the differing levels of SCM maturity among supply network members.

No two providers possess exactly the same technical or managerial capabilities. An approach based on the use of design chains recognises this and the inability of traditionally defined work packages to accommodate the technical differentials between organisations that is a product of the increasing specialisation of knowledge in design. The use of design chains requires a different approach to defining work packages:

- Providers should be allowed to contribute to a receiver's definition so that the provider's view of what is strategically important can be fully considered by the receiver.

- The use of 'roles and responsibility matrices' (also known as demarcation schedules or scoping agreements) provide the basis for collaboratively developing typical work packages or templates for work packages. Applying Design Management Practices (PT1), Performing an ADePT Review (BT6) and Applying ADePT to Design Management (PO1) give details as to when the matrices might be needed and can be used in conjunction with generic design process maps (see Modelling Business Design Processes (BO1) and Applying Supply Chain Management in the Business (BT3)).

Providers

- In fully developed supply networks, providers should be permitted to, and should be capable of, proposing the content of their own work packages because they should be developing a detailed understanding of their receiver's requirements through their evolving business and project relationships. Once proposed by a provider, the actual content can be agreed by negotiation between the provider, the receiver and the providers of other work packages that are linked to the one under discussion. This ensures that a mutually acceptable allocation of tasks results that involves an appropriate compromising in the drawing of boundaries between organisations (recognising that an optimal solution is not possible as different organisations will always have different working methods and objectives, irrespective of their membership of a supply network).

Receivers

- Receivers must recognise that they are incapable of fully defining the scope of design responsibility without involving other design chain members. This is because allocating design responsibility needs to be done on the basis of work-package definition and, for this, input from the providers is needed.

- Receivers need an in-depth knowledge of other organisation's capabilities, including their SCM maturity before they can accurately define a work package. This can be gained through auditing the supply network (see Auditing the Supply Network (BO2)).

- Receivers must focus on structuring work packages to suit what is best for the project rather than what is best for them.

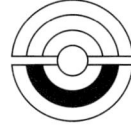

Establishing Project Ground Rules

The activities to be performed within a project need to be reviewed at the outset by all project parties. This establishes a shared understanding of the project and of the ground rules that must be adopted to ensure that the practices, which need to be performed, get done efficiently.

These rules define who does what within the design chain (task allocation) and how tasks are performed (process definition). On a project-by-project basis, they define the role each organisation needs to take and their interaction with other project parties. Ground rules may also address additional issues, such as the manner by which partnering principles will be applied to a project.

There is an opportunity for all members of the design chain to discuss the rules (by Assembling Supply Chains (PO4)) to ensure that the rules retain legitimacy so that all organisations buy-into them and are willing to adopt them in their practices.

Ground rules can assist project parties that wish to partner with each other (by Planning Supply Chain Management Project Practice (PS2)), but have not yet established a shared understanding of how to go about it. In part, the required understanding of partnering agreements develops over time; however, if the ground rules can be agreed beforehand, the process can be shortened. Such ground rules might include:

- procedures for agreeing the management of a project programme;
- procedures for introducing new individuals and late-appointed organisations into a project;
- methods of reducing 'churn' in project personnel in order to preserve the common value system and the way ground rules are understood; and
- the use of project charters or protocols to express project ground rules - these need to be backed-up by agreed working methods developed by the project team members.

These mechanisms all contribute to other areas of practice, including developing buy-in to projects, team building and building consensus on projects (see Assembling Supply Chains (PO4) for more information).

Selecting Communication Mechanisms

Communication is central to effective design. Many mechanisms and issues surrounding SCM address this issue, such as building consensus, team building, the use of stage gates, launch meetings and co-location. Effective communication within a supply chain can also be achieved using:

- project newsletters, which can be used to keep all project parties informed of latest developments;
- 'redline' meetings, which bring designers together in appropriate groups to resolve outstanding issues;
- 'principals meetings', which bring senior individuals together to discuss overall project issues, such as the effectiveness of current project procedures or other areas of concern; and
- project web-sites and project intranets,[1] which provide useful centralised information resources that can be rapidly and easily updated - the need to routinely review these should be included in the induction and ongoing education of all project members.

[1] Faraj, I., Alshawi, M., Aouad, G., Child, T. and Underwood, J. (2000) An Industry Foundation Web-based Collaborative Construction Computer Environment: WISPER, *Automation in Construction*, 10, pp. 79-99; and Baldwin, A., Thorpe, A. and Carter C. (1999) The Use of Electronic Information Exchange on Construction Alliance Projects, *Automation in Construction* 8, pp. 651-662.

Using Team Building Techniques

Team building (usually formalised) is used to establish a sense of mutual understanding, respect and co-operation 'for the good of the project'. Team building can instil team spirit in all project members, irrespective of the organisation to which they belong. Creating project environments, in which everyone pulls together, facilitates the social relationships necessary to integrate design chain activities. By giving project objectives priority over organisational and personal ones, the achievement of the best possible project solutions is promoted. It also ensures that the project is not subject to the sort of compromises that might otherwise be made by individual members of the design chain if they were working on their own.

When building project teams, careful attention must be paid to both the reporting structures and to the organisation boundaries. It is important to balance the informal working practices that arise between individuals (and the organisations they represent) with the formal mechanisms that govern relationships between organisations. This is to ensure an appropriate balance of flexibility and expediency with the need for rigorous, auditable processes.

In design chains, the complexity of the interfaces between organisations always requires some accommodation within the informal working methods, to overcome the inevitable shortcomings of formal project management structures. This sort of accommodation is facilitated by team building, which itself is greatly assisted by project working methods that use ground rules that have been established by consensus (see Assembling Supply Chains (PO4)).

When building project teams, careful attention should be given to the following:

- Involving appropriate people. If team-building workshops are held during the early project stages, they will tend to involve only people at the higher management levels. The team works best when it is focused on those involved in the day-to-day running of a project.

- The need for team-building workshops to be focused must be balanced with the need for a certain amount of socialising to reinforce working relationships.

- Using the project as the focus for the team. One way this might be achieved is by holding a technical workshop within the project environment. This sort of focus permits role swapping, facilitated team activity, and other collaborative, team-based approaches. These may be required to resolve complex technical issues unsuited to traditional methods, such as technical queries.

- Ensuring team building is genuine and is capable of producing constructive criticism, such as questioning the suitability of the formal processes. At the same time, while some formal training in team-building techniques is useful (the teaching of listening skills, for example), this must be put in the context of the individual projects so that the relevance of the techniques are grounded in the practical problems facing the project team.

- Team building as a continuous exercise, lasting for the duration of the project. All appropriate opportunities for team building should be exploited (such as bringing together the design teams from different organisations or from different departments of the same organisation).

- Accelerating team building by establishing core teams to bring individuals and organisations together on a project to create a cadre of experience around which further teams can form and accelerate team building. The core team acts as a social framework of experience and established relationships, and others then absorb these values.

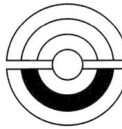

PT3

Selecting Supply Chain Members at the Project Level

Provider benefits

- For the provider the benefit of this selection process is the emphasis that is placed on the delivery of value and service to the project.

- Providers will know that they will have been selected on the basis of their ability rather than for any other reason - such as the lowest bid.

- A transparent selection process can help the provider understand the reasons why it was selected (or not selected) for a particular project, and this can help the provider plan how it updates its skills and market appeal.

Receiver benefits

- The receiver benefits by being able to allocate workload on the basis of the project needs for service and value, with price being a product of business relationship.

- This selection process encourages consistency in supply chain assembly. The selection of the supply chain is justified to the client on the basis of project criteria, thereby avoiding recourse to competitive tendering.

Purpose

To make use of the technical design competencies that exist within the supply network. A supply (and, therefore, design) chain, as discussed earlier, is formed by a group of organisations, each one of which has one or more specialised skills they can contribute to projects. Together, the chain combines a broader range of technical competencies than is likely to be found in any one organisation. The tactic of selecting supply chain members for particular projects, and how they can be drawn from business supply networks into the project supply chain, is clearly important and is what is discussed here. This project tactic uses information provided in Auditing the Supply Network (BO3).

Summary

A supply chain is assembled from a business supply network and is built by matching organisations with particular technical skills and SCM maturity to project requirements. The supply network may exist informally (as an individual firm's preferred business partners) or as a managed network linked by long-term business relationships and/or by formal agreements.

The selection of supply chain members requires a deeper understanding of a company's capabilities and competencies than is usually required by traditional arrangements. This understanding must address technical expertise, the capacity to perform design and the management ability to integrate the firm's competencies with other organisations.

The selection process needs to consider a number of factors about potential supply chain members. It might look at:

- project criteria - to ensure appropriate organisations are appointed to suit technical project characteristics;

- whether there is a mismatch between an organisation's competencies and its capabilities (the difference between the theoretical ability to be involved in a project and having the available resources to contribute at a particular moment in time);

- suitable sources of information - collected through the supply network or past performance on projects - about the potential supply chain members; and

- the gathering of descriptions of organisations (derived by Auditing the Supply Network (BO3)) to select supply chain members.

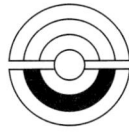

Outline Procedure

Several factors can contribute towards the way supply chain members are selected:

1. Project criteria
2. Using assessment data
3. Competencies and capabilities
4. Drawing together appropriate information

Requirements and Resources

- Senior project team members, including project managers, design managers and principal designers, are responsible for assembling the project supply chain. To complete their work, they require access to generic design and work package definitions, and the information provided by value surveys and supply network audits.

Related Practices

- ☐ **BS1** Planning Design Process Management
- ☐ **BS2** Planning Supply Chain Management Business Practice
- ☐ **BS3** Planning the Implementation of Integral Value Engineering across the Business

- ☐ **BT1** Applying Process Management in the Business
- ☑ **BT2** Aligning Supply Networks
- ☐ **BT3** Applying Supply Chain Management in the Business
- ☐ **BT4** Applying Integral Value Engineering in the Business
- ☑ **BT5** Conducting a Value Survey
- ☐ **BT6** Performing an ADePT Review

- ☑ **BO1** Modelling Business Design Processes
- ☑ **BO2** Auditing the Supply Network
- ☐ **BO3** Auditing the Supply Network for Technical Competence
- ☑ **BO4** Gathering Value-adding Feedback from Projects

- ☐ **PS1** Planning Project Design Management
- ☐ **PS2** Planning Supply Chain Management Project Practice
- ☑ **PS3** Planning Integral Value Engineering Project Practice

- ☐ **PT1** Applying Design Management Practices
- ☐ **PT2** Applying Supply Chain Management to a Project
- ☐ **PT3** Selecting Supply Chain Members at the Project Level
- ☑ **PT4** Implementing Integral Value Engineering on a Project

- ☐ **PO1** Applying ADePT to Design Management
- ☑ **PO2** Applying DePlan to Design Management
- ☐ **PO3** Modelling Project Design Processes
- ☑ **PO4** Assembling Supply Chains
- ☑ **PO5** Applying Value-adding Tools to Design Problems

Project Criteria

As discussed above, supply chain members are selected, in part, on the basis of project criteria. This is usually determined by the receiver, although sometimes it is done by the client. Where competitive tendering occurs, project criteria can be the basis by which providers go on the shortlist.

The receiver sets the criteria by looking at the key project drivers. These can be identified using value management in the traditional manner. Their identification can be complemented by using providers from the supply network as a resource to provide specialised advice, even if they are no longer used on the individual projects. If this is the case, the selection of the final design chain should formalise their involvement, such as through payment for early design advice.

Project criteria covers areas such as:

- expected scope of design involvement;
- expected involvement in value management;
- core technical design capabilities required;
- core management skills required;
- minimum performance levels (e.g. quality, risk management, delivery, health and safety, etc.);

- level of SCM maturity required (including particular practices for a project);
- anticipated value of the works;
- anticipated time scale; and
- geographic location of the project.

Providers

- As part of the on-going dialogue within the supply network, providers will increasingly be seen as a resource for receivers to approach, in the early development stages, for consultation on the project and for testing project criteria throughout the project. The benefits of the membership of the supply network would act as a check on either party taking advantage of their business relationships.

Receivers

- Determining the project criteria provides an effective way of crystallising project-team thinking. This activity may share many characteristics with Planning Supply Chain Management Project Practice (PS3).

Using Assessment Data

Once collected, assessment data are mapped against the project criteria to build a picture of the available resources within the supply network and those that must be provided by the project supply chain. Using techniques such as gap analysis, or by plotting available skills against standard roles and responsibility matrices, receivers can assess the suitability of particular providers. Done in this way, the selection process can be seen to have been both fair and rigorous; and can be opened to scrutiny by one-off clients.

Receivers

- In less mature supply networks, a receiver will perform this process either in isolation or with the client. As supply network maturity develops, and the business relationships within it deepen, the management of the system will change and will become less authoritarian - the receiver choosing the providers and in a more consensual process - where the supply network as a whole selects the project supply chain.

Competencies and Capabilities

Competence is the ability of an organisation to perform an activity. In the design chain, competency is concerned with an organisation's ability to design, manage its design activities and integrate them with other supply chain members. Capability is an organisation's capacity to perform a particular activity with competence. With project criteria determined, the selection of supply chain members then becomes influenced by matching required competencies with available capability in the supply network.

The supply network provides a forum for continuous discussion between providers and receivers. This helps each to understand the other's current and expected workload and allows them to warn each other when particular resources - such as designers - are in high or low demand. If provider or a receiver finds that its current workload prevents it having the capability required to competently undertake a design task, then other supply network members may provide that capability.

Providers

- Discussion in the supply network keeps providers and receivers abreast of developments so that steps can be taken to balance resource supply. This puts providers in a better position to withstand pressure from receivers to take on work which exceeds their capacity.

Receivers

- By balancing design responsibilities with the capacity to deliver, receivers can create a project supply chain in which each member can deliver what is expected of it. This increases management confidence and leads to tighter planning and reduced risk on future projects.

Drawing Together Appropriate Information

The information required to assemble a project supply chain is drawn from a variety of sources. Some of this information is project specific while other information will be generated from within the business domain. The information a receiver might require (based on project criteria) includes the following:

The expected scope of the design involvement: This is drawn from the generic process maps and generic work packages. These are described by Modelling Business Design Processes (BO1) and Applying Integral Value Engineering in the Business (BT4).

The expected provider involvement in value management and integral value engineering: This can be gathered by a value survey. The way to do this is discussed in Conducting a Value Survey (BT5).

The core technical capabilities required (the technical ability to provide a design solution): In part, the core technical capabilities are identified by models of providers' business processes (see Modelling Business Design Processes (BO1)) but useful information can also be obtained by reviewing the technical skills identified by Auditing the Supply Network for Technical Competence (BO3) and by Auditing the Supply Network (BO2) and Assembling Supply Chains (PO4).

The core management skills required: Details of these are available from a variety of sources, including the receiver's records of providers, and past project performance records, which are a part of the project reviews generated in the supply network (see Assembling Supply Chains (PO4) and Auditing the Supply Network (BO2)).

The minimum performance levels: Receivers assess issues of time, cost, quality, value, risk, health and safety when appointing providers. Using the technical assessments detailed in Auditing the Supply Network (BO2) and Auditing the Supply Network for Technical Competence (BO3) project benchmarks can be established to ensure a minimum level of performance for the supply chain.

The level of SCM maturity required: This can be assessed by conducting a supply network audit (see Auditing the Supply Network (BO2)).

PT4

Implementing Integral Value Engineering on a Project

Provider benefits

- Provides access to IVE resources, such as a value-adding toolbox, to all organisations involved in projects, irrespective of whether they are maintained with their own business domain or that of a receiver with whom they are working in a project design chain.

Purpose

The objective is to define working methods for individual projects based on the IVE resources provided by the organisations operating in the design chain associated with a project. This practice is closely associated with that of Applying Integral Value Engineering in the Business (BT4).

Summary

This practice extends the deployment of the IVE resources identified in the business domain to projects. Ideally, these resources will be managed by a single organisation on behalf of a project design chain (including a shared or rented toolbox). In this situation, it is essential that all other organisations have appropriate IVE resources deployed and their staff trained. The task is to identify those required on the project and make them available to all members of the design chain and encourage their use. When IVE is applied unilaterally by a single organisation, no further tactical responses are required beyond that in the associated business tactic (BT4).

Receiver benefits

- Ensures that IVE resources deployed in the business are applied to projects in an effective manner, maximising the collaborative working on design problems and the feedback to the business and future projects.

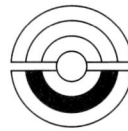

Outline Procedure

The steps in applying IVE to projects are:
1. Determine results required
2. Plan and develop approaches
3. Deploy approaches
4. Assess and review approaches

Requirements and Resources

- Planning Integral Value Engineering Project Practice (PS3) defines how IVE resources are identified by business tactics and introduced to project activity by this practice.

Related Practices

☐ **BS1** Planning Design Process Management

☐ **BS2** Planning Supply Chain Management Business Practice

☐ **BS3** Planning the Implementation of Integral Value Engineering across the Business

☐ **BT1** Applying Process Management in the Business

☐ **BT2** Aligning Supply Networks

☐ **BT3** Applying Supply Chain Management in the Business

☑ **BT4** Applying Integral Value Engineering in the Business

☐ **BT5** Conducting a Value Survey

☐ **BT6** Performing an ADePT Review

☐ **BO1** Modelling Business Design Processes

☐ **BO2** Auditing the Supply Network

☐ **BO3** Auditing the Supply Network for Technical Competence

☐ **BO4** Gathering Value-adding Feedback from Projects

☐ **PS1** Planning Project Design Management

☐ **PS2** Planning Supply Chain Management Project Practice

☑ **PS3** Planning Integral Value Engineering Project Practice

☐ **PT1** Applying Design Management Practices

☐ **PT2** Applying Supply Chain Management to a Project

☐ **PT3** Selecting Supply Chain Members at the Project Level

☐ **PT4** Implementing Integral Value Engineering on a Project

☐ **PO1** Applying ADePT to Design Management

☐ **PO2** Applying DePlan to Design Management

☐ **PO3** Modelling Project Design Processes

☐ **PO4** Assembling Supply Chains

☐ **PO5** Applying Value-adding Tools to Design Problems

Determine Results Required

The results required from IVE on a project determines how IVE resources should be made available for use. In particular, the mechanisms used to disseminate these resources throughout the project environment must not comprise their ability to:

- support the achievement of strategic goals - this includes gathering information from everyday activities to inform the strategic direction of IVE on the project; and
- make resources available for use in the project in a manner that will allow its strategic goals to be met without impeding ongoing design activity.

The implementation of IVE on a project must, therefore, be planned to determine the information requirements of the strategies supported by this tactic, in addition to providing ease of access to IVE resources for project members.

The results required from the tactical application of IVE (and IVE resources) to a project will depend on the nature of the project, the circumstance in which it is undertaken, the organisations associated with it, and the IVE resources available for application. Example outcomes typically sought from planning IVE practice on a project include:

- establishing a consistent understanding of value and IVE across the project team;
- supporting problem-solving in design, as this can be particularly insightful when involving providers and receivers from more than one design chain organisation;
- encouraging individuals and teams within the design chain to undertake IVE collaboratively;
- documenting evidence of value-adding activity, for reference by end-user customers and within IVE practice on future projects;
- support IVE by providing software tools; and
- gathering feedback on IVE resources and providing this to the business domain to inform their maintenance and development.

Plan and Develop Approaches

Appropriate IVE resources must be selected for the project in hand, including both tools and techniques, and the education and training required to use them. The choice will be influenced by the scale of implementation and the proficiency of those involved (both at an organisational and individual level). Ideally this should be done at the project outset and integrated into working methods and design management. This maximises availablity and minimises resistance to change. Education and training is essential and best achieved through a workshop involving all organisations. This encourages shared understanding, common goals and helps to identify and overcome barriers. It is also an opportunity for individuals to identify new IVE techniques for the toolbox.

Deploy Approaches

The selected IVE resources must be deployed from their source organisation into all other members of the design chain. This will require access to the value-adding toolbox through the internet or a project extranet. See Applying Integral Value Engineering in the Business (BT4).

Providers

- Opportunity to learn new IVE techniques and introduce in-house methods into the project and to other providers.

Assess and Review Approaches

It is necessary to extend resource use monitoring in projects into the business domain to some degree, so that design chain organisations can learn to improve their IVE practice. This involves gathering examples of IVE resource use from the current project, archiving them in the business domain, for dissemination into future projects, where the same circumstances arise, and this evidence would form useful examples. It is vital that evidence of IVE resource use is gathered as projects progress, as it is much more difficult to capture lessons after a project has concluded.

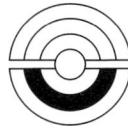

Applying ADePT to Design Management

Provider benefits

- Improved understanding of the information transfer between design chain organisations.
- Improved co-ordination at design chain interfaces.
- Gives clear guidance on critical decision-making actions.
- Gives the provider a basis to employ tactics to resolve design process bottlenecks.

Purpose

By applying the Analytical Design Planning Technique[1] (ADePT) to design management, organisations can identify the optimal sequence of design activities to satisfy the development of a design solution, thus creating an opportunity to develop detailed design programmes that take into account the iteration inherent within the design process. By using the business and project design process models, receivers and providers can use ADePT to analyse the complex interfaces that exist between organisations on projects.

Summary

ADePT consists of three main components:

- a model of the design process defining supply network activities and their information requirements,
- a dependency structure matrix (DSM) analysis which is linked to the model and identifies the optimal sequence of tasks and iteration within the design process; and
- a design programme that is integrated with the project plan through further DSM analysis.

The technique acknowledges the iteration in design and identifies and programmes multidisciplinary and multi-organisational co-ordination activities. Unlike traditional network analysis techniques, ADePT represents iteration in the design process allowing the relevant design and management tasks to be accurately programmed.

Receiver benefits

- Improved understanding of the provider's position within the design process.
- Illustrates the impact of receiver decision making, e.g., timeliness of the delivery of design information, and the effect on the design programme.
- Improved integration and programming of detailed design .
- Allows the receiver to pre-define group decision-making opportunities to resolve bottlenecks in the design process.

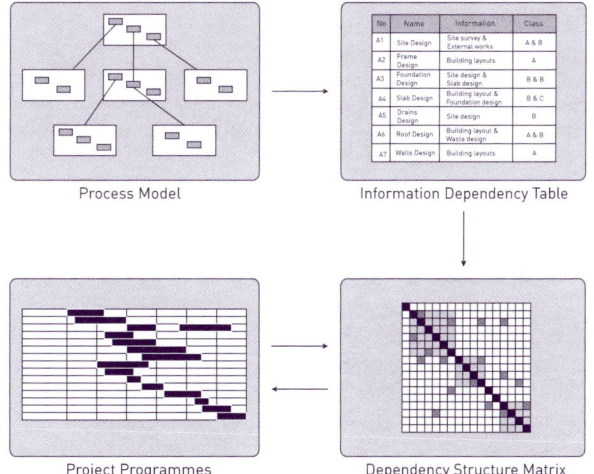

Process Model

Information Dependency Table

Project Programmes

Dependency Structure Matrix

1 See www.adeptmanagement.com for more information on ADePT and its related software.

Outline Procedure

ADePT can be applied to design management by:
- Tailoring the business design process model to represent a project specific design process model (refer to Modelling Business Design Processes (BO1) and Modelling Project Design Processes (PO3)).
- Classifying information dependencies.
- Analysing the dependency structure matrix (DSM).
- Developing a design programme (by adding activity durations and resource constraints).
- Creating an integrated project programme.

Requirements and Resources

- The ADePT methodology must be available to all organisations involved in a project environment so that they can analyse business and project design processes, and plan design activity (see www.adeptmanagement.com). By using ADePT, organisations can view the complex relationships that exist internally and externally, thus providing them with the opportunity to enhance organisational learning.
- Project process models must be available to the team for project application (see Modelling Project Design Processes (PO3)).
- Process modelling software for maintaining graphical models.
- DSM analysis software for optimising design processes and integrating with construction.
- Project management software to assist in the production of design project programmes.

Related Practices

- ☐ **BS1** Planning Design Process Management
- ☐ **BS2** Planning Supply Chain Management Business Practice
- ☐ **BS3** Planning the Implementation of Integral Value Engineering across the Business

- ☐ **BT1** Applying Process Management in the Business
- ☐ **BT2** Aligning Supply Networks
- ☐ **BT3** Applying Supply Chain Management in the Business
- ☐ **BT4** Applying Integral Value Engineering in the Business
- ☐ **BT5** Conducting a Value Survey
- ☐ **BT6** Performing an ADePT Review

- ☑ **BO1** Modelling Business Design Processes
- ☐ **BO2** Auditing the Supply Network
- ☐ **BO3** Auditing the Supply Network for Technical Competence
- ☐ **BO4** Gathering Value-adding Feedback from Projects

- ☐ **PS1** Planning Project Design Management
- ☐ **PS2** Planning Supply Chain Management Project Practice
- ☐ **PS3** Planning Integral Value Engineering Project Practice

- ☐ **PT1** Applying Design Management Practices
- ☐ **PT2** Applying Supply Chain Management to a Project
- ☐ **PT3** Selecting Supply Chain Members at the Project Level
- ☐ **PT4** Implementing Integral Value Engineering on a Project

- ☐ **PO1** Applying ADePT to Design Management
- ☐ **PO2** Applying DePlan to Design Management
- ☑ **PO3** Modelling Project Design Processes
- ☐ **PO4** Assembling Supply Chains
- ☐ **PO5** Applying Value-adding Tools to Design Problems

Classifying Information Dependencies

Design decisions have a varying impact on the design process. Some are fundamental to the delivery and construction of a design solution, others are incidental. ADePT employs a classification system to represent this issue and is made on the basis of three factors:

- strength of dependency of the task on the information;
- sensitivity of the task to slight changes in the information; and
- the ease with which the information can be estimated.

Information flow	Task is	Task is	Information is
Class A	INCREASINGLY DEPENDENT ↑	INCREASINGLY SENSITIVE ↑	INCREASINGLY ESTIMABLE ↓
Class B			
Class C			

Some classifications may be generic, and hence applicable to all projects, whereas others may be specific to the nature of the project.

Providers

- Providers should seek an understanding of how their information is to be used by receivers - to inform the qualitative definition of the operation.

Receivers

- The receiver is responsible for analysing the importance of each information interdependency and assigning a suitable classification.

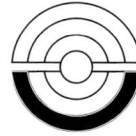

Analysing the Dependency Structure Matrix (DSM)

A DSM is a matrix analysis approach that uses an algorithm to sequence tasks in an optimised order that minimises iteration. This iteration can be kept to a minimum, and introduced only where it is necessary to ensure the design process has been explored fully. The purpose of optimising a matrix is to maximise the availability of information required, and minimise the amount of unnecessary iteration and the scale of concurrent design activity within the blocks. The tasks are then scheduled to minimise the number of dependencies above the diagonal of the matrix.

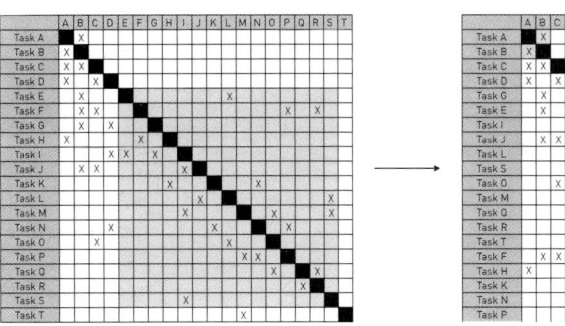

Optimising a matrix sequences activities that do not contribute to design blocks. It identifies those within design blocks, but does not sequence the activities within the blocks. The activities within a block should be scheduled to suit the most appropriate mode of concurrent working.

In some cases it is necessary to reduce the size of a block of concurrent design tasks so that they can be planned and managed more effectively. This is called tearing, and involves minimising feedbacks by either identifying estimates that can be made with confidence or by fixing a design parameter. When doing this, the design team are negotiating and making decisions to declassify specific information dependencies.

Providers

- Providers must ensure that declassifications are feasible.

Receivers

- It is important that providers of design solutions are fully engaged in this process and the subsequent management of the project. All designers should understand their position in the overall process and be actively encouraged to preempt possible problems.

Developing a Design Programme

The output from an optimised DSM must be represented on a programme in a manner that incorporates the iteration within the process. This is done by assigning durations and resources, and outputting the DSM sequence into a programme. It will be necessary to group tasks that form a design block under a 'rolled up' activity, removing interrelationships from the block. Each block needs to be considered in its own right and programmed in such a way that both initial and final co-ordination can be achieved, resources are allocated according to their availability and the overall project duration is achieved.

While the resulting design programme will look the same as a conventional bar chart, it is fundamentally different in that the analysis has taken account of all information flows, including feedbacks. Standard project management tools based on the critical path method, can not represent iteration.

A12	Site Design		A	A			B
A252	Road & Car Park Design	A		A			
A13	Primary Elements Design						B
A251	External Works Design	A	B	C			
A233	Retaining Walls Design	C		B	C		
A1311	Basement GA					A	

ID	Name		Nov	Dec	Jan
24	Site & External Work Design Exercise				
25	A12	Site Design			
26	A252	Road & Car Park Design			
27	A13	Primary Elements Design			
28	A251	External Works Design			
29	A233	Retaining Walls Design			
30	A1311	Basement GA			

Providers

- Providers should assess durations for activities and ensure that these reflect any streamlining of the design process.

Receivers

- Receivers should plan for sign-off at the end of each iterative block of design activities to avoid abortive work.

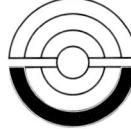

Creating an Integrated Project Programme

In order that a project can be planned with ADePT, the optimal design programme must be integrated with the procurement and construction programmes. The timing of involvement of the design chain organisations will vary according to the stage of the project, and the design programme needs to take this into account. This can be achieved by either changing the duration of some tasks, with corresponding allocation of resources, or by changing the sequence of tasks through the estimation or fixing of some information. An analysis of the effects of changing the sequence of tasks is undertaken through an examination of the process on a DSM, highlighting the need for iteration between the DSM and programming stages of ADePT.

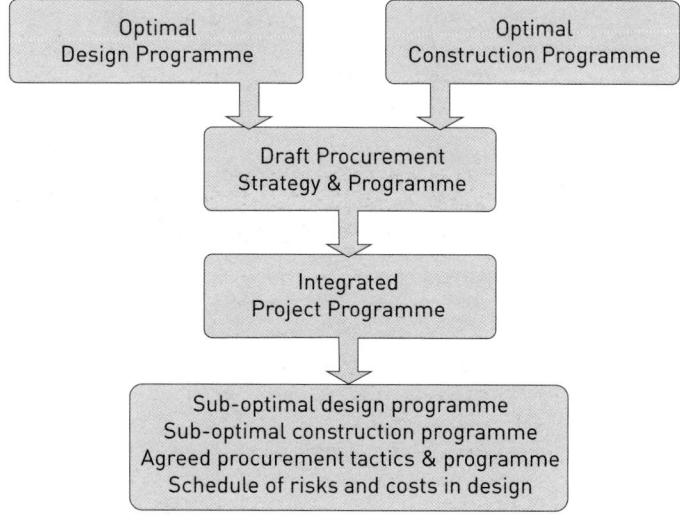

Optimal Design Programme

Optimal Construction Programme

Draft Procurement Strategy & Programme

Integrated Project Programme

Sub-optimal design programme
Sub-optimal construction programme
Agreed procurement tactics & programme
Schedule of risks and costs in design

Receivers

- In the past design programmes have been over-constrained to suit construction, with the consequence of many design problems arising during construction. ICD recognises this and encourages greater attention to the need for properly programmed design that should reduce abortive work.

Applying DePlan to Design Management

Provider benefits

- By using DePlan, a provider will know what information other members of the design chain require. This will ensure that the provider can focus upon the timely delivery of critical pieces of design information and, at the same time, remain committed to accomplishing design activities on programme.

- The transparent nature of DePlan ensures that the provider can take account of the design information demands of the design chain, thus a provider can be confident that they are producing information that is required by others.

- Improves performance to programme and provides a measure of percentage planned complete (PPC).

Receiver benefits

- Using DePlan to plan and control the design process will allow receivers to be certain that information is delivered to them from other design chain organisations to enable them to undertake their own design activities.

- A receiver can monitor a provider to determine what design information is outstanding, thus allowing the receiver to understand barriers to progress, and replan to ensure success.

- By not undertaking design that does not have all the necessary pieces of information available, the receiver can make sure they do not produce abortive design output.

Purpose

By applying DePlan to design management practices, a design chain can be sure that the design project is undertaken in a transparent environment, where the design chain organisations are aligned, and information transfer is streamlined. Project design activities can be completed on programme by improving the planning and control of task scheduling with a focus on removing design activity constraints and only starting activities that have a high likelihood of timely completion.

Summary

DePlan is a design management tool that provides design chain organisations with a means to plan strategically for iteration and to schedule work in a realistic and controlled manner, leading to improved design process reliability and certainty. DePlan ensures that the design chain can work in an integrated way to deliver a design solution. The cross-organisational interfaces are managed by employing a project design process model to identify design information constraints hindering the successful completion of design activities. This benefits the project team because the project design process has exactly the same underlying logic used to develop the project design programme (see Applying ADePT to Design Management (P01)). Hence, applying DePlan to design management practice offers the design chain a consistent approach to planning and managing the design process.

The figure below illustrates the components of DePlan, where design is planned using ADePT, which also provides the team with a set of project design process models (see Modelling Project Design Processes (P03)) that can be used to identify design information constraints acting on planned design activities. Short-term planning is a detailed view of the overall design plan, focused on those tasks scheduled for completion within the planning window. Once the planned activities have been undertaken, a performance review reports the percentage planned completed (PPC). This review will provide feedback to the design chain, indicating what has been achieved, and the reasons for failure for those that were not completed.

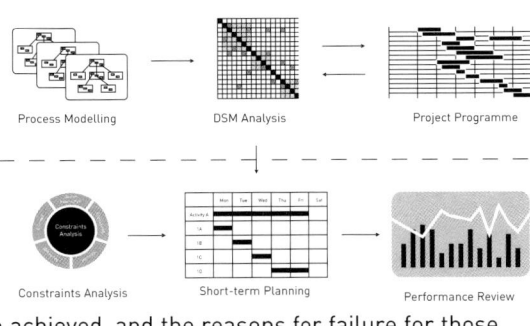

Process Modelling   DSM Analysis   Project Programme

Constraints Analysis   Short-term Planning   Performance Review

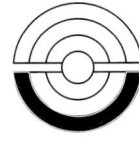

Outline Procedure

DePlan can be applied to design management by:
- Producing the project design programme (this is achieved by Applying ADePT to Design Management (PO1)).
- Producing a short-term plan
- Analysing design activity constraints
- Producing a weekly design activity programme
- Conducting a design activity performance review

Requirements and Resources

- It is important that all design chain organisations use DePlan to manage their design process, so that transparency is achieved and maintained. Therefore, DePlan must be made available to all organisations, and suitable training provided to its users.
- Receivers must be responsible for managing and administrating the use of DePlan. It is important for receivers to understand and communicate their requirement to providers.
- Providers have an onus to provide the relevant information to the receiver when requested. If this is not achievable, the provider must inform the receiver and alternative arrangements be made for information delivery.
- Receivers will need to use a spreadsheet to calculate the PPC, when determining performance.

Related Practices

- ☐ **BS1** Planning Design Process Management
- ☐ **BS2** Planning Supply Chain Management Business Practice
- ☐ **BS3** Planning the Implementation of Integral Value Engineering across the Business

- ☐ **BT1** Applying Process Management in the Business
- ☐ **BT2** Aligning Supply Networks
- ☐ **BT3** Applying Supply Chain Management in the Business
- ☐ **BT4** Applying Integral Value Engineering in the Business
- ☐ **BT5** Conducting a Value Survey
- ☐ **BT6** Performing an ADePT Review

- ☐ **BO1** Modelling Business Design Processes
- ☐ **BO2** Auditing the Supply Network
- ☐ **BO3** Auditing the Supply Network for Technical Competence
- ☐ **BO4** Gathering Value-adding Feedback from Projects

- ☐ **PS1** Planning Project Design Management
- ☐ **PS2** Planning Supply Chain Management Project Practice
- ☐ **PS3** Planning Integral Value Engineering Project Practice

- ☐ **PT1** Applying Design Management Practices
- ☐ **PT2** Applying Supply Chain Management to a Project
- ☐ **PT3** Selecting Supply Chain Members at the Project Level
- ☐ **PT4** Implementing Integral Value Engineering on a Project

- ☑ **PO1** Applying ADePT to Design Management
- ☐ **PO2** Applying DePlan to Design Management
- ☐ **PO3** Modelling Project Design Processes
- ☐ **PO4** Assembling Supply Chains
- ☑ **PO5** Applying Value-adding Tools to Design Problems

Producing a Short-term Plan

The design programme developed by applying ADePT to design management practices represents the whole design project plan, and for the purposes of applying DePlan to design management, this programme should be viewed in the short-term so that the design chain can focus upon the completion of critical design activities. The short-term plan is a view of the planning window, with design activities shown at the detail level.

Providers

- The provider is responsible for identifying their design activities that exist within the short-term plan.
- The provider must identify all design activities in detail, ensuring that they follow the priority dictated by the project design programme.

Analysing Design Activity Constraints

The design activities represented by the short-term plan should be reviewed by the receiver to identify all their information requirements, and these must be communicated to the provider, allowing suitable time to ensure timely delivery. Also, other constraints should be identified, such as: resource requirements, contractual conditions, materials and engineering needs. The receiver is also responsible for identifying these constraints, and communicating them to the provider. Once all design activity constraints have been identified, designers must work together to ensure that all constraints are removed prior to weekly planning of design activities.

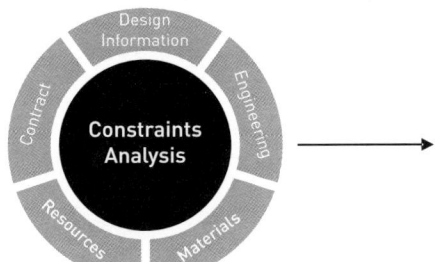

Activity: Plant Floor General Arrangement Design			
Information Required	Disc	Due	Complete
Plant Room Prim Member Layouts	Struct	1st Sept	Y/N
Plinth & Plant Bases Details	Arch	1st Sept	Y/N
Plant Room Floor Calcs	Struct	1st Sept	Y/N
Primary Elements Layouts	Arch	1st Sept	Y/N

Providers

- The provider must ensure that all requested pieces of design information are delivered to the receiver by the specified time. If there is a risk of failing to deliver, the provider must inform the receiver so that the dependent design activity can be re-planned, thus ensuring that design activity is not undertaken if the likelihood of failure is high.
- The provider must analyse all activities within the short-term plan to identify design activity constraints.
- The provider must monitor the progress of the design activity constraints against the priority of the short-term plan, ensuring that the critical design activities are ready for weekly planning when dictated by the design programme.

Receivers

- The receiver is responsible for communicating their needs to the provider, allowing sufficient time for the provider to deliver the necessary information prior to the programmed commencement of the design activity.

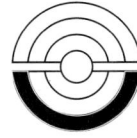

Producing a Weekly Design Activity Programme

The constraint-free design activities shown on the short-term plan must be planned on a weekly basis, following the priority as shown on the design programme. Because the activities have been reviewed, and the constraints identified and subsequently removed, the likelihood of successful completion should be high. Therefore, a receiver and a provider can be confident that the design process is moving forward. By planning design activities in the short-term, design teams can focus on the priority tasks, thus ensuring that the design programme is being adhered to. Should there be any circumstances where activities within the short-term cannot be completed (due to too many constraints) then they should be re-planned and the design programme amended accordingly.

Providers

- The provider must be sure that all known design activity constraints are removed prior to design activity weekly planning
- The planned design activities must conform to the priority as shown on the design project programme

Receivers

- Receivers should only plan to achieve those activities for which the design constraints have been removed.

Conducting a Design Activity Performance Review

Once the planned design activities have been undertaken, the performance of the design chain should be reviewed. The provider must review the weekly plan to determine which design activities have been completed successfully, and those that have failed to be achieved. If activities are completed, then the provider must ensure that the project design programme is updated to reflect progress. This performance review is called percentage planned complete (PPC) and involves calculating the percentage of design activities completed from the weekly planned set. The activities that were not completed must be reviewed to determine what reasons led to the failure. These reasons, plus an outline description describing the measures to ensure future success must be fed back to the design chain, so that the project design programme can be re-planned.

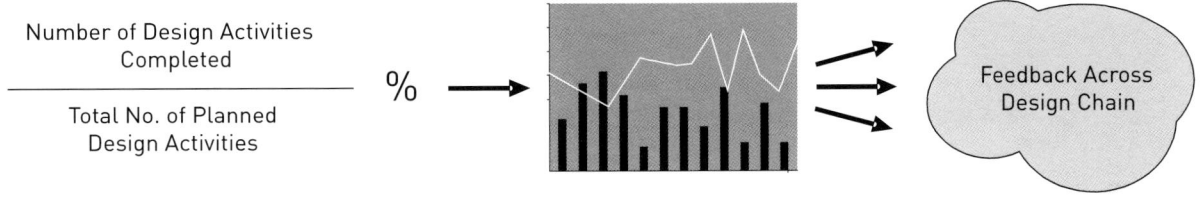

Providers

- The provider is responsible for reviewing the performance of design activities at the end of each planned week.
- Incomplete activities must be reviewed by the provider to determine the cause of failure.
- All incomplete activities must be re-planned, ensuring that the reasons for failure have been overcome.

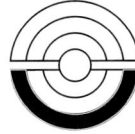

P03

Modelling Project Design Processes

Receiver benefits

- Improved understanding of what information is required by the receiver.
- Provides mechanism for clarification of what information is to be provided, leading to the efficient co-ordination of design information with the receiver.
- Improved planning of design information transfer.

Provider benefits

- The establishment of a project process model enables the receiver to coherently communicate their information requirements to supporting providers.
- By understanding the process, the receiver can optimise the sequence of design activities, based upon the flow of information internally and from supporting providers (if done using ADePT, see Applying ADePT to Design Management (P01)).
- Ensures the smooth transfer of information flow to and from providers.
- Ensures that the receiver is not replicating design work done by the provider.

Purpose

By modelling project design processes, the design operations of organisations can be represented in a simple way and aligned in the project design chain. This involves selecting appropriate parts of business design process models to represent the design chain members' contribution to a project. The production of a project design process model will ensure any overlaps or gaps in the design scope of the project design chain will be avoided. The definition of the information flowing throughout the project will increase transparency in the design chain and enable the integration, planning and management of design activity.

Summary

The project process model is the project specific representation of design activities taken from the business models, which represent each ICD organisation (see Modelling Business Design Processes (B01)). The organisational models are reviewed against the project brief and schematic design to determine the systems and components that constitute the product. Each organisation then tailors their individual model(s), where required, to represent their specific contribution. Consequently, the detailed level activities and internal/external information flows in the business process models need to be reviewed and amended.

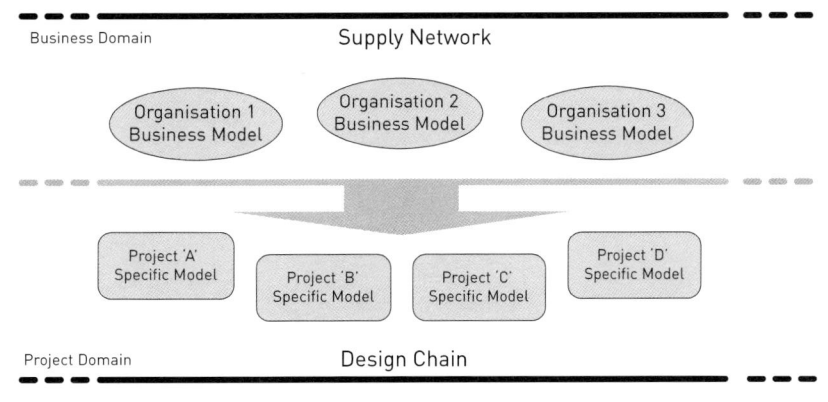

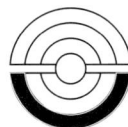

Outline Procedure

A project design process model can be developed by:

1. Defining the project design process work breakdown structure
2. Identifying gaps and overlaps
3. Aligning business and project design processes
4. Feedback learning to business process models

Requirements and Resources

- Before developing a process model, the providers and receivers involved must adopt the ICD principle of viewing projects as processes. This is required to support definition of their processes, leading to the integration, planning, and management of design by Applying Design Management Practices (PT1) and Applying ADePT to Design Management (PO1).
- The receiver must have followed Modelling Business Design Processes (BO1) to define the design activities that are reviewed to determine the project processes.
- Receivers are responsible for communicating their information needs to providers when developing project process models. This will ensure that the provider does not prescribe the receiver/provider relationship, as its definition is best led by the receiver, given that it must co-ordinate its interfaces with many concurrent providers.
- When developing project process models, the receiver must ensure that sufficiently knowledgeable and skilled individuals are involved.

Related Practices

- ✔ **BS1** Planning Design Process Management
- ☐ **BS2** Planning Supply Chain Management Business Practice
- ☐ **BS3** Planning the Implementation of Integral Value Engineering across the Business

- ☐ **BT1** Applying Process Management in the Business
- ☐ **BT2** Aligning Supply Networks
- ☐ **BT3** Applying Supply Chain Management in the Business
- ☐ **BT4** Applying Integral Value Engineering in the Business
- ☐ **BT5** Conducting a Value Survey
- ☐ **BT6** Performing an ADePT Review

- ✔ **BO1** Modelling Business Design Processes
- ☐ **BO2** Auditing the Supply Network
- ☐ **BO3** Auditing the Supply Network for Technical Competence
- ☐ **BO4** Gathering Value-adding Feedback from Projects

- ☐ **PS1** Planning Project Design Management
- ☐ **PS2** Planning Supply Chain Management Project Practice
- ☐ **PS3** Planning Integral Value Engineering Project Practice

- ✔ **PT1** Applying Design Management Practices
- ☐ **PT2** Applying Supply Chain Management to a Project
- ☐ **PT3** Selecting Supply Chain Members at the Project Level
- ☐ **PT4** Implementing Integral Value Engineering on a Project

- ✔ **PO1** Applying ADePT to Design Management
- ☐ **PO2** Applying DePlan to Design Management
- ☐ **PO3** Modelling Project Design Processes
- ☐ **PO4** Assembling Supply Chains
- ☐ **PO5** Applying Value-adding Tools to Design Problems

Defining the Project Design Process Work Breakdown Structure

Design decisions have a varying impact on the design process. Some are fundamental to the delivery and construction of a design solution, others are incidental. ADePT employs a classification system to represent this issue and are made on the basis of three factors:

- strength of dependency of the task on the information;
- sensitivity of the task to slight changes in the information; and
- the ease with which the information can be estimated.

Some classifications may be generic, and hence applicable to all projects, whereas others may be specific to the nature of the project.

Roles and responsibilities for design activity can be developed from the work breakdown structure, and this can be communicated to all members of the design chain in the form of a matrix (CIRIA has provided guidance to assist in this). This provides the design team with a clear indication of who is responsible for which design activities, thus ensuring that each organisation is allocated a design role that suits their capability.

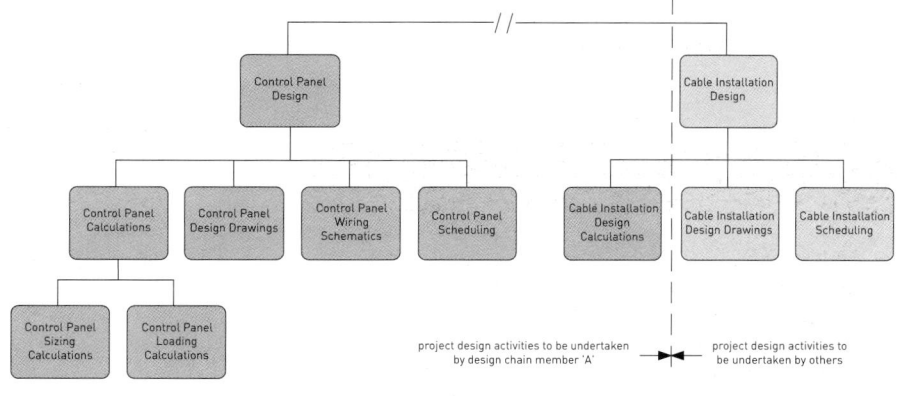

	1. General Design Activity			
Design Activity	Responsibility (please tick)			Additional Explanation
	Designer	Installer	Other	
Control Panel Calculations				
Control Panel Sizing Calcs Control Panel Loading Calcs				
Control Panel Design Dwgs				
Control Panel Wiring Schematics				
Control Panel Scheduling				
Cable Installation Design Calcs				
Cable Installation Design Dwgs				
Cable Installation Scheduling				

Receivers

- The receiver is responsible for ensuring that the cross-organisational information flows satisfy their information needs.

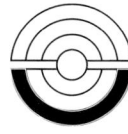

Identifying Gaps and Overlaps

There may be instances where the business design process models representing the providers within the design chain either do not cover the technical requirements of the design project or replicate activity. These gaps and overlaps must be identified and removed by introducing additional skills where necessary and allocating duplicated capabilities to the appropriate party. This is achieved by aligning the business and project design processes.

This provides a response to Latham's observation[1] that the UK construction industry needs to clearly understand the allocation of design responsibilities between design chain members if project design activity is to be effective.

Aligning Business and Project Design Processes

The project work breakdown structure is used to define the design scope of the construction project. The collaborating organisations within the design chain must review the project process structure to align technical capability (represented by their business design process model) and the project design process. The cross-organisational flow of information should be used to define the interfaces between organisations. These interfaces are critical to the successful alignments of business and project design processes, and the subsequent planning and management of design activity.

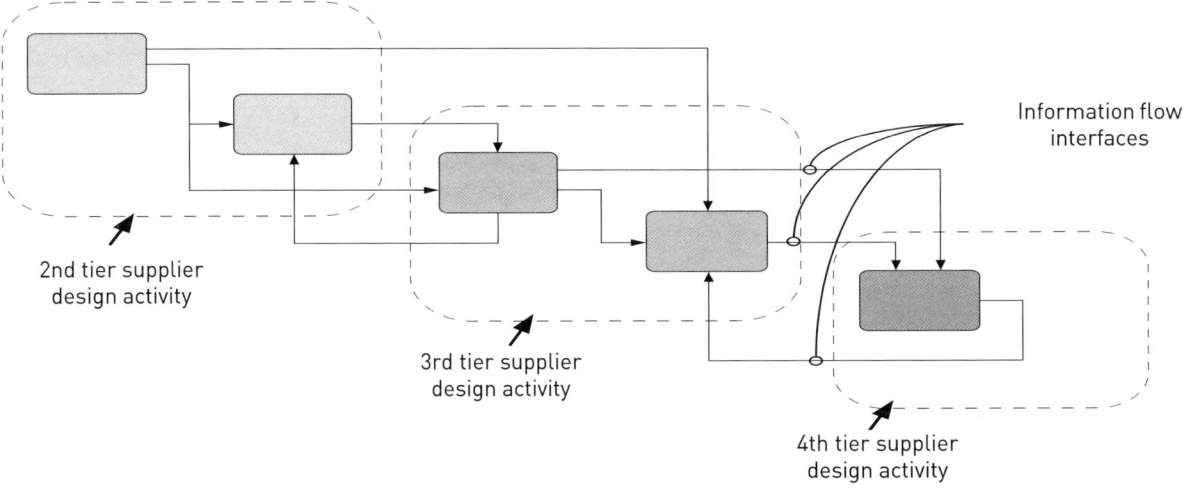

2nd tier supplier
design activity

3rd tier supplier
design activity

4th tier supplier
design activity

Information flow
interfaces

Providers

- The provider must review the project structure and identify those areas that match their technical capability through negotiation with other members of the design chain. This negotiation is best overseen by a third party to ensure that design responsibility is allocated to improve the combined design ability of the design chain, and does not benefit one design chain member at the expense of others.

[1] Latham, M. (1994) *Constructing the Team: Joint Review of Procurement and Contractual Arrangements in the United Kingdom Construction Industry*, HMSO, London.

Feedback Learning to Business Process Models

Having applied design process models to the project environment, organisations should feedback experiences from the application of these models to the business domain so that any learning can be captured and incorporated within the generic models. This review should reflect upon both the internal processes and the key interfaces with other organisations. Where appropriate, the feedback should be passed on to appropriate members of the supply network. This will ensure that each ICD organisation can improve its performance within the supply network based upon the learning while engaged as part of a design chain.

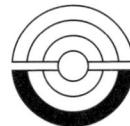

P04

Assembling Supply Chains

Purpose

Once organisations have been selected for inclusion in a supply network and their competencies have been identified, they can be formed into a design chain to structure their involvement in projects. This practice gives advice describing how this can be done and how design chains can be used to address some key areas of collaborative working: procurement, incentivisation, innovation, and communication.

Summary

This practice is associated with the practices: Planning Supply Chain Management Project Practice (PS2) and Applying Supply Chain Management to a Project (PT2). It describes a variety of techniques that providers and receivers can use to form a project supply chain from a business domain supply network. These techniques include:

- mechanisms to build consensus on projects - to align understanding of the project objectives and the expectations of each organisation;
- technical clusters - grouping organisations together according to the interdependence of their activities on projects;
- co-location of design teams - to improve communication and facilitate team building;
- project reviews - to identify any lessons from projects; and
- launch and 'stage-gate' meetings - creating opportunities to review past performance and communicate issues to personnel across organisations.

Provider benefits

- Through techniques such as building a consensus, taking advantage of technical clusters and 'stage-gate' reviews, providers can gain greater influence over project management decisions made by the receiver than would otherwise be the case.

- Projects run more smoothly if participants (collaborators or parties) can share project objectives and agree on the strategy to achieve them - through consensus-building mechanisms; this is a factor that applies equally to providers and receivers.

Receiver benefits

- For receivers, projects generally work best when the providers are fully committed to what they are doing. The techniques detailed represent ways of binding providers into the project through non-contractual methods.

- Technical clusters allow receivers to delegate much of the management between providers in the design chain to the participants in the value chain.

- Projects run more smoothly if participants (collaborators or parties) can share project objectives and agree on the strategy to achieve them - through consensus-building mechanisms; this is a factor that applies equally to providers and receivers.

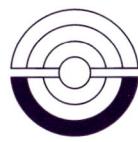

Outline Procedure

There is no set order for these procedures. We have found the order given below works well but, depending upon circumstance, other approaches might be just as effective.

1. Developing consensus in the supply chain
2. Technical clustering of providers
3. Design team co-location
4. Project reviews
5. Launch and stage-gate meetings
6. Responding to innovation

Requirements and Resources

- Many of the techniques identified in this practice are applicable on projects where SCM is not formally being applied. However, in order to gain the full benefits of these techniques, this practice should follow on from Planning Supply Chain Management Project Practice (PS2) and Applying Supply Chain Management to a Project (PT2). No specialist personnel are required beyond the normal complement of project staff, although deployment of additional resources, such as facilitators for team building exercises, may be effective.

Related Practices

- [ ] **BS1** Planning Design Process Management
- [ ] **BS2** Planning Supply Chain Management Business Practice
- [ ] **BS3** Planning the Implementation of Integral Value Engineering across the Business

- [✓] **BT1** Applying Process Management in the Business
- [✓] **BT2** Aligning Supply Networks
- [ ] **BT3** Applying Supply Chain Management in the Business
- [✓] **BT4** Applying Integral Value Engineering in the Business
- [ ] **BT5** Conducting a Value Survey
- [ ] **BT6** Performing an ADePT Review

- [✓] **BO1** Modelling Business Design Processes
- [✓] **BO2** Auditing the Supply Network
- [ ] **BO3** Auditing the Supply Network for Technical Competence
- [ ] **BO4** Gathering Value-adding Feedback from Projects

- [ ] **PS1** Planning Project Design Management
- [✓] **PS2** Planning Supply Chain Management Project Practice
- [✓] **PS3** Planning Integral Value Engineering Project Practice

- [ ] **PT1** Applying Design Management Practices
- [✓] **PT2** Applying Supply Chain Management to a Project
- [ ] **PT3** Selecting Supply Chain Members at the Project Level
- [ ] **PT4** Implementing Integral Value Engineering on a Project

- [✓] **PO1** Applying ADePT to Design Management
- [ ] **PO2** Applying DePlan to Design Management
- [ ] **PO3** Modelling Project Design Processes
- [ ] **PO4** Assembling Supply Chains
- [ ] **PO5** Applying Value-adding Tools to Design Problems

Developing Consensus in the Supply Chain

The ICD approach to building consensus on projects (which overlaps with the 'project partnering' aspect of the Planning Supply Chain Management Project Practice (PS2)) is about developing and communicating a shared understanding of project aims and objectives. This is to ensure that all design chain members 'buy-in' to what everyone is trying to achieve. The intention is to have the design chain function as a single entity, thereby aligning project objectives.

Agreement of project aims and objectives among all members of the design chain is particularly important when the chain comprises a large number of organisations; the further an organisation is from the client, the more it relies on the rest of the design chain to communicate the project's aims and objectives. Parties with aligned objectives will find it easier to assess added value. Organisations that do not share the same project objectives effectively only respond to clearly defined criteria (such as the need to reduce cost). In contrast, the greater benefits of IVE (for example) are achieved though resolving complex problems; projects may be delivered to budget but they are unlikely to exceed client expectations.

There are a number of techniques that can be used to build consensus and to align project objectives; as a rule, the earlier they are implemented the better they work.

- Value management and value engineering. This can be used to structure and agree project objectives. Value management forums are particularly effective with the client being involved in setting the design chain objectives.

- Project briefings. A formal project 'kick-off' meeting provides the opportunity to communicate to all project designers some of the context surrounding the project. This increases the likelihood of them signing up to a common project strategy. It also provides a forum for the formal communication of a strategic project plan within the design chain, which may include defining how the various tools and techniques used by ICD will be deployed in the project.

- Toolbox talks. Originally developed as a site practice, where workers were kept up-to-date with the project, informal design-orientated 'toolbox talks' can explain the project content to designers to give them a wider context of their design in relation to other designers in the design chain. These talks help designers from different organisations (or different parts of a large organisation) appreciate their interdependency and helps them understand how their activities impact on others (e.g. bringing the information in design process maps to life).

- Published project strategy documents. By making a project strategy document available to all design chain members, all are in a position to understand the rationale of how the work has been structured; if people do not understand, they can, at least, ask questions. Many organisations already create such documentation but usually it is maintained for internal purposes only and is seldom shared between collaborating organisations.

Providers

- Providers benefit from a shared understanding of the project's objectives; it puts them in a better position to appreciate the factors which will affect a receiver's decision making and, thus, they can more easily predict how a receiver might respond to change.

- The value of consultation between provider and receiver also increases when it is clear that both see a project in the same way. The provider, rather than just reacting to a strategy set out by the receiver, gets an opportunity to contribute directly.

Receivers

- A benefit for receivers in being able to work effectively with providers in developing project strategies is that this prevents providers from later claiming ignorance about aspects of the project's intent; it also encourages them to be better prepared when key decisions need to be taken.

- When a consensus has been achieved, and providers accept and begin to contribute to the project, the team has the opportunity to start testing the logic and the assumptions of the decisions they have made. Without a consensus, this type of review can become distorted with different parties needing to 'fight their own corners'. With a consensus, the exercise can provide valuable early warning of problems yet to be encountered.

Technical Clustering of Providers

When design responsibilities are allocated within projects, organisations that need to work closely with each other can be brought together to form clusters. Such clusters might be between all those concerned with the building envelope, such as the structural designer, architect, steelwork subcontractor, cladding, roofing and curtain walling specialist contractors. It is useful to identify such groups because these are the areas were problems of information flow are most likely to occur within the design chain. The techniques described in the practices Modelling Business Design Processes (BO1) and Applying ADePT to Design Management (PO1) provide a basis for this to be done.

Clusters work most effectively if a leader (who is usually, but may not necessarily be the receiver) is appointed. The leader has the responsibility of managing the design within the cluster. However, selecting the right leader is not necessarily a straightforward task as they need to have the appropriate skills. Organisation characteristics, such as SCM maturity and managerial skills, can be used as a way of doing this (details of how these characteristics can be identified are given in Auditing the Supply Network (BO2)).

Providers

- Clusters enable providers to take control of the relationships between themselves and to co-ordinate their work schedules and the flow of information. Taking ownership enables them to have the appropriate flexibility within the constraints set on the cluster by the receiver. It means they can identify more closely with what they are doing. Thereby, providing benefits in the management of design and organisational interfaces, and in the allocation of risk.

Receivers

- Clustering enables receivers to delegate the management of inter-organisational interfaces, within strictly controlled remits, to providers. It eliminates the need to micromanage providers and provides a structured framework for providers to interact with one another without losing overall control of the provision of design solutions and allows effective risk allocation between the members of the design chain.

[1] See: Holti, R., Noicolini, D., Smalley, M. (2000) *The Handbook of Supply Chain Management*, Tavistock Institute, CIRIA, London.

Design Team Co-location

Co-locating - having designers physically near to each other - often comes about as a result of clustering. Co-location can take many forms: providers can move to receivers' offices, or vice-versa, or both providers and receivers can move to a new dedicated location, such as a project office. Co-location can also include several tiers of a design chain including the client. The benefits of co-location include:

- improved communication (particularly informal communication);
- one point of contact for the client;
- faster decision making achieved through the immediacy of contact (e.g. coffee pot meetings);
- improved team building, achieved by developing a common project culture and identity that is associated with the sharing of physical space - this is particularly beneficial when individual designers are working away from home; and
- a fast-track method of integrating the provider's and receiver's processes.

Despite these benefits, the practice of co-location creates other issues that need to be taken into account:

- Co-location may require some individuals to work away from home. For office-based designers this may not be desired. Management must therefore recognise this 'home/base' issue when assessing the benefits of co-location and the utilisation of their overheads.
- An organisation's input on the project may not justify full-time attendance by their personnel. Individuals within smaller organisations are likely to be working on several projects simultaneously. In these situations, co-location may have a detrimental effect on other projects in which they are involved and may not be cost-effective.
- If co-location causes one party to be seen as 'playing at home' while others are seen to be 'playing away', the discomforted party may see the arrangement as unfair and this can affect the smooth running of the project. Careful management can iron out these sorts of issues but it is imperative to recognise the potential problems.

Some of these issues are beginning to be addressed through the development of collaborative technologies, such as GroupWare, instant messaging and video conferencing. Development in virtual teamworking may reduce some of the need to physically co-locate while still facilitating the personal contact necessary for collaborative working.

Providers

- The scale of individual projects and the size of the organisations involved will influence the ability of organisations to respond to the co-location of design staff.

Receivers

- Co-location can provide an opportunity for closer or tighter control of the project through the physical proximity of staff from different organisations.

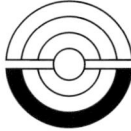

Project Reviews

Project reviews are important in their own right and have a strategic role in assembling the supply chain. The performance of individual organisations in the design chain, in part and collectively, can be assessed with recommendations made for future projects. Reviews are typically carried out in a workshop environment and involve a range of project personnel from senior management to individual project designers. Project reviews can be carried out in a number of ways:

- A one-off review can be performed at a project's conclusion and can examine all the activity that took place. However, the complexity of some projects, and the number of changes that might have taken place during the project - in the organisations and with the personnel involved - can mean that such reviews can be too unwieldy to be useful.

- Periodic reviews are a way of avoiding the loss of the detail in the all-encompassing end-of-project review. These can also throw up useful lessons more quickly and can be run in conjunction with stage gates.

- A targeted review at the outset of a new project creates an opportunity to gather all project members together to examine their collective knowledge, gathered from their previous projects. This allows practices that have been used effectively by at least one supply chain member in the past to be replicated throughout the project supply chain, bringing its benefit to the new project.

In SCM, the involvement of both receivers and providers in project reviews is critical to identify the important lessons that need to be fed back to the business domain. The process of learning from projects and project members, capturing knowledge and making it available to the network, is an important part of Planning the Implementation of Integral Value Engineering across the Business (BS3), Aligning Supply Networks (BT2) and Applying Integral Value Engineering in the Business (BT4).

Providers

- Finding the resources necessary to hold project reviews can often be a factor that prevents them from taking place. It is the end-of-project review that suffers particularly. The need to fully utilise resources means that project staff are often quickly moved on to new projects when old projects finish and it becomes difficult to reform the original teams to carry out a review.

Receivers

- Finding the resources necessary to hold project reviews can often be a factor that prevents them from taking place. It is the end-of-project review that suffers particularly. The need to fully utilise resources means that project staff are often quickly moved on to new projects when old projects finish and it becomes difficult to reform the original teams to carry out a review

Launch and Stage-Gate Meetings

Stage-gates are reviews held at key stages during a project, such as the project launch and the construction launch. They bring different organisations, or different parts of the same organisation, together. A stage-gate is a point where, because all the important players are present, important decisions can be made, reviews can be held and outstanding issues resolved. It is also the point at which forthcoming work can be planned, especially as this may need to take account of critical decisions that might have just been made.

Stage-gates act as both a review of the previous 'stage' of the project (see Applying Process Management in the Business (BT1)) and as launch point for the next stage. As such, they are usually work as a 'review board' for the key relevant stakeholders. This can then be combined or followed with a meeting of the people involved in the next stage.

Stage-gate meetings are also useful opportunities for managing and for increasing the consensus within a project (and a 'consensus-building' launch meeting discussed above is a specialised form of stage-gate meeting). They can be used to get buy-in from the design chain, to develop a shared understanding of project objectives and examine how the project is progressing. For large firms, where design takes place as a distinct activity within the organisation, stage-gate meetings can be particularly useful because they can be used as a mechanism to align the different parts of a business and to overcome functional silos that may have developed.

Providers

- Both providers and receivers may be involved in the review boards or launch meetings.

Receivers

- Both providers and receivers may be involved in the review boards or launch meetings.

---

[2] Cooper, R.G. (1990). 'Stage-Gate System: A New Tool for Managing New Products', *Business Horizons*, May-June, pp. 44-54, see also the stage-gate approach in the Process Protocol and Figure 2.8.

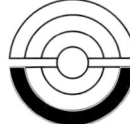

Responding to Innovation

Without ICD, the relationship between providers and receivers is often unbalanced. Traditionally innovation in working methods has been introduced in a 'top-down approach'. With ICD, the need for receivers to procure design expertise from providers means such receivers tend to be more receptive to project innovations suggested by the providers. At the same time, providers themselves need to be alive to other innovative ideas that might come from others in the design chain.

In ICD, innovative ideas are able to move easily from the project domain to the business domain because of the emphasis given to improving communication between the two. If a receiver becomes aware of new techniques a provider has pioneered, the receiver, working through the supply network, can apply the new techniques to any of its current projects or those it might undertake in the future. Similarly, innovation instigated by a receiver can be just as easily communicated to providers, and to other receivers on other design chains, by the same mechanism.

Maximising the potential for innovation that providers can bring to projects requires careful management. On occasion providers may come up with the ideas, but if they think they will not be involved in the parts of the project where the new ideas may be applied, they may be unwilling to share them. Being clear about what the provider is capable of is one way of resolving this issue (see Planning Integral Value Engineering Project Practice (PS3)).

Another way of addressing this sort of problem is by rewards. Providers need to be paid for their innovative thinking. There is a danger that the payment may not fully match the value of the new idea, but at least the receiver, in making the payment, is able to send a signal. Alternatively, a formal method of allocating the costs and benefits of innovation can be agreed between supply chain members at a project's outset. This may be in terms of a percentage of the savings made as a result of the innovation or a share of a pooled bonus, which accrues to all members of the supply or design chain for good performance. This latter technique can be particularly effective because it encourages innovations that assist the whole project, rather than individual work packages or organisations.

Receivers

- Receivers must capture, consider and respond to product or process innovations suggested by providers during the course of a project so that those innovations may be applied to those parts of the project still to be completed.

Applying Value-adding Tools to Design Problems

Provider benefits

- Value-adding tools help structure the identification and dissemination of best practice. The more providers work with the tools, the more they can refine effective ways of working and thus ensure these processes can be applied to subsequent work. Where these processes integrate the design input of the provider with that of other providers and receivers, they help secure involvement in subsequent projects and sustain long-term business relationships.

Receiver benefits

- The assembly of value-adding tools into problem-solving processes creates opportunities to 'step back' from ongoing design activity and consider how the design tasks add value to a project. Documenting this reflective exercise can provide evidence of value-for-money of the design.

- The use of value-adding tools to structure problem solving in design adds rigour to the process, and provide additional justification for deploying the technical solutions that satisfy project values.

Purpose

Value-adding tools need to be applied to design problems within a problem-solving framework that delivers the appropriate project values. This ensures that designers work with the design chain to deliver value.

Summary

Problem-solving techniques are assembled from a toolbox of value-adding tools to create a framework that addresses specific design tasks or groups of tasks (see Performing an ADePT Review (BT6)). Project values should be communicated to the team and defined as additional design criteria. Successful approaches should be reused in other solutions, and the way it is done is documented so it can in included in the value-adding tools when they are next reviewed (by Gathering Value-adding Feedback from Projects (BO4)). By doing this, organisations can maintain a portfolio of value-adding tools and include lessons learned, helping designers repeat effective solutions or avoid repeating those found to be unsuccessful. A greater range of relevant experiences will be available if the toolbox is a shared resource (see Implementing Integral Value Engineering on a Project (PT4)).

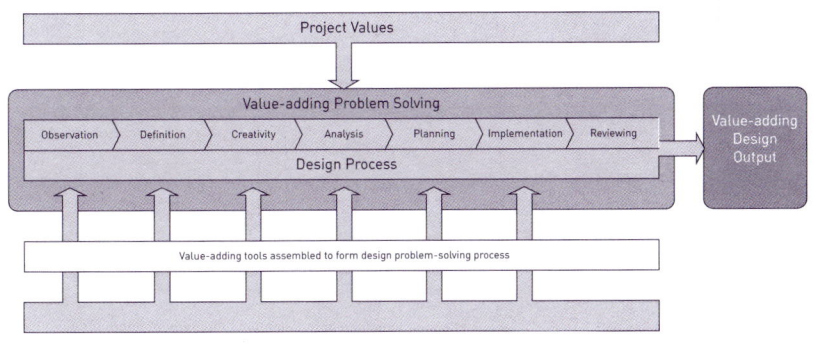

Outline Procedure

Value-adding tools can be applied to design problems by:
1. Defining design problems
2. Selecting suitable tools and assembling a problem-solving process
3. Documenting specific applications

Requirements and Resources

- A value-adding toolbox must be made available to all organisations involved in a project so that they can learn from the IVE lessons it embodies. This toolbox may be operated by a single project party but, to share these insights among collaborating organisations, it must be maintained as a common resource in either a design chain or value system.

Related Practices

- [ ] **BS1** Planning Design Process Management
- [ ] **BS2** Planning Supply Chain Management Business Practice
- [ ] **BS3** Planning the Implementation of Integral Value Engineering across the Business

- [ ] **BT1** Applying Process Management in the Business
- [ ] **BT2** Aligning Supply Networks
- [ ] **BT3** Applying Supply Chain Management in the Business
- [ ] **BT4** Applying Integral Value Engineering in the Business
- [ ] **BT5** Conducting a Value Survey
- [x] **BT6** Performing an ADePT Review

- [ ] **BO1** Modelling Business Design Processes
- [ ] **BO2** Auditing the Supply Network
- [ ] **BO3** Auditing the Supply Network for Technical Competence
- [x] **BO4** Gathering Value-adding Feedback from Projects

- [ ] **PS1** Planning Project Design Management
- [ ] **PS2** Planning Supply Chain Management Project Practice
- [ ] **PS3** Planning Integral Value Engineering Project Practice

- [ ] **PT1** Applying Design Management Practices
- [ ] **PT2** Applying Supply Chain Management to a Project
- [ ] **PT3** Selecting Supply Chain Members at the Project Level
- [x] **PT4** Implementing Integral Value Engineering on a Project

- [ ] **PO1** Applying ADePT to Design Management
- [ ] **PO2** Applying DePlan to Design Management
- [ ] **PO3** Modelling Project Design Processes
- [ ] **PO4** Assembling Supply Chains
- [ ] **PO5** Applying Value-adding Tools to Design Problems

Defining Design Problems

The scope of the problem must be defined. At later stages of the project, IVE allows individual technical design solutions to be managed as individual problem-solving processes. Within an ICD approach, IVE is applied continuously during all project stages where design takes place.

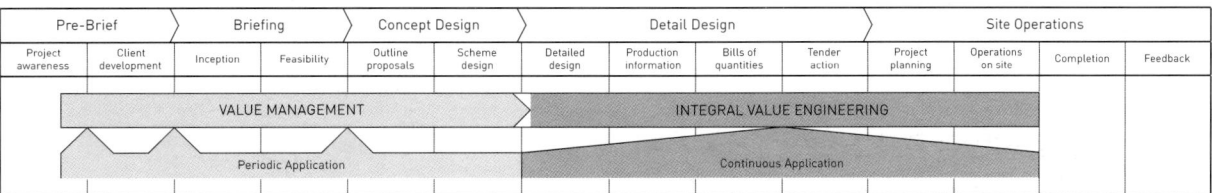

Pre-Brief		Briefing		Concept Design		Detail Design						Site Operations		
Project awareness	Client development	Inception	Feasibility	Outline proposals	Scheme design	Detailed design	Production information	Bills of quantities	Tender action	Project planning	Operations on site	Completion	Feedback

VALUE MANAGEMENT — INTEGRAL VALUE ENGINEERING

Periodic Application — Continuous Application

This allows the design problems to be structured with value-adding tools and undertaken as a series of focused problems. Generally, problems become smaller and more specific as projects progress.

Value-adding tools can address both problem definition and resolution. This extends the variety of tools used in IVE beyond those currently deployed in value management.

Selecting Suitable Tools and Assembling a Problem-solving Process

The effectiveness of value-adding tools relies on the richness of the toolbox from which they have been drawn. It contains the tool descriptions and guidance on how to use them. Numerous approaches are applied to help designers select tools suited to their needs:

1. The stage of the project at which the design problem considered occurs.
2. The stage of the problem-solving process.
3. The technical characteristics of the design problem.
4. Key words or phrases contained in tool descriptions.
5. A list of all value-adding tools available in the toolbox.

Once a number of tools have been selected, they are assembled into a problem-solving process and technical solutions developed (using appropriate design methods) within this framework, with value-adding tools complementing that activity. Examples of value-adding tools include: Function Analysis; FAST Diagramming; Affinity Diagramming; Value Trees; SCAMPER; Force Field Analysis; and the Needs-Metrics Matrix. These have been gathered from a variety of sources including established value management practice, general management problem-solving and product development in the manufacturing sector.

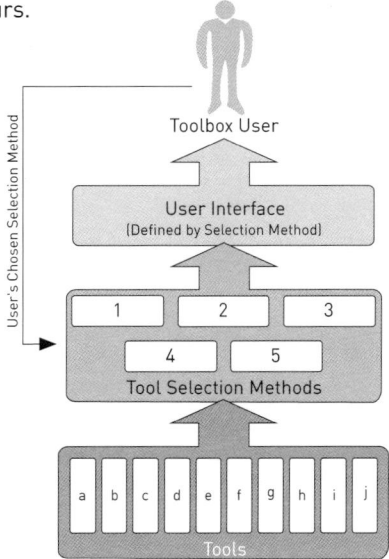

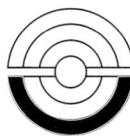

Documenting Specific Applications

Value-adding tools, help relate the technical implications of design decisions to project values and explicitly demonstrate how the solution responds to the project values. These typically comprise functional objectives (or simple statements of requirements) for the whole project, derived from value management exercises carried out earlier in the project. Alternatively, if the technical problem is sufficiently important within the context of the whole project, more specific values may be identified (with project client and stakeholder input) for it to satisfy. The use of value-adding tools in this way can demonstrate rigorously the attainment of value.

Applications should be documented in a standard format. These records form part of a value-adding toolbox and inform subsequent selection of its tools, either elsewhere on the project or on future projects. The standard format should describe the overall problem-solving process and tool set before detailing how each tool was applied. This documentation forms a vital pat of the feedback to the business domain.

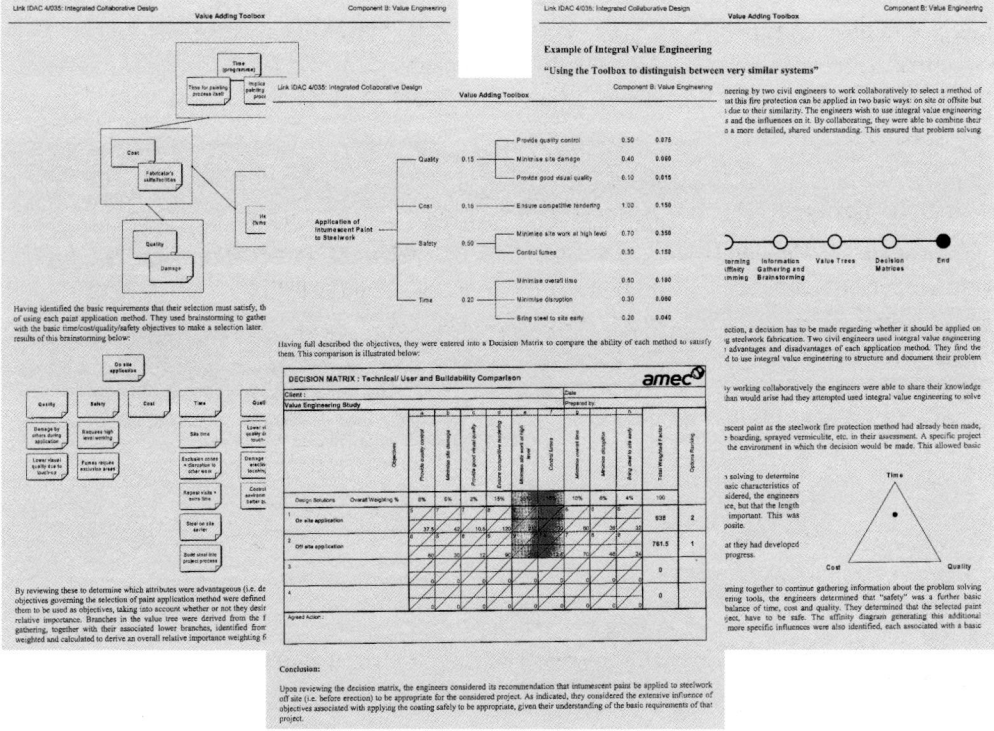

Glossary

Business
The permanent organisation which exists to provide the infrastructure to bring resources together to provide goods and services and manage the delivery of successive projects.

Business Domain
The long-term environment in which an organisation deploys strategies and develops structures to its revise working methods, amend its culture and manage its alignment with other permanent organisations. Members of a design chain co-ordinate their long-term business development strategies to develop mutual strategies and align their operations and cultures within the business domains of their respective organisations.

Capability
An organisation's physical ability to provide a technical competency. Capability is determined by a number of technical factors including, for example, an organisation's geographical coverage and the numbers of design and management staff it has available to resource a project at any point in time.

Client
The client is the organisation or individual that requires the functionality of the completed project procured from the construction industry. This operational need must justify each construction project undertaken.

Competency
The technical ability to provide a product or service to another party within a supply chain. Within a design chain, the commodity of exchange is design information. The competency is the technical ability arising from the combinations of resources within an organisation but is not related to the extent of resource available.

Design
The act of devising solutions to the problem of satisfying end-user/client requirements through the provision of the project output. Within an ICD approach, 'design' represents the act of designing within the design chain, rather than the documented evidence of the design solution. If divided into smaller processes to ease management, individual design processes resolve design processes of less scope. Collectively, however, their output remains the same.

Design Block
A group of interdependent tasks, identified in a Dependency Structure Matrix (DSM) as a shaded block of activities.

Design Chain
That aspect of project supply chains that is managed and co-ordinated to benefit its members and project clients through improved co-ordination of design processes.

Design Information
The documented output of design processes, containing the specialised knowledge contributed by each design chain member to the technical solutions they collaboratively produce.

Design Information Flow
The transfer of design information from one design process to another, representing its progression from one stage in its development to the next. Within a design chain, information can flow between design processes performed within a single chain member, or it can flow across the interfaces between organisation interfaces.

Design Process
An element of design activity that exclusively processes design information generated by preceding design processes to generate further design information.

Design processes are not commenced until all the information they require is available, allowing them to progress without interruption. Depending on their scope, they typically exist within a single design discipline and are usually undertaken by a single project designer. Design processes occur on projects, but are defined and allocated in the business relationships of design chain members.

Design Process Model (DPM)

A simple representation of the activities undertaken by the supply network organisations during the design phase of the construction project. The model represents activities and their interrelationship through the flow of design information.

Dependency Structure Matrix (DSM)

A matrix analysis approach that uses an algorithm to sequence sets of interdependent tasks in an attempt to make information available to complete a task as and when it is required.

Domains

All actions performed by a single organisation occur within either its business domain or its project domain. These define the environment within which an organisation is concerned at the time of implementing its business development measures.

Optimisation

The act of optimising the sequence of design processes within a project programme according to the flow of information between them. This is an automated procedure, performed by the software supporting the Analytical Design Planning Technique (ADePT).

Integral Value Engineering (IVE)

The continuous consideration of value throughout project design, irrespective of the stage of project progression at which the design takes place, the designers involved, or the technical systems considered. IVE is a core element of an ICD approach to design management and, once established, lies within the core values and attitudes of all design chain members, underpinning their basic approach and desire to complete design in a value-adding manner. It provides the basis for managing design activity to ensure that it results in solutions that add value to the client.

Integrated Collaborative Design (ICD)

An approach to managing design information production within projects that establishes design as the common thread linking organisations together to harmonise the interfaces and working practices.

IVE Resources

The varied working practices or tools used in the business or project domain to support and assist an organisation, value system or design chain in its performance of Integral Value Engineering.

Key Aspect

One of a series of common principles identified as being fundamental to supply chain management.

Market

All the available potential suppliers of products and services within the construction industry, known or unknown to any organisation.

Maturity

The level of sophistication with which an organisation plans and manages its interfaces with other supply chain members.

Organisation

The permanent economic endeavour that employs resources and possesses competencies for application to individual construction projects

Operations

The everyday activities of the organisations that have adopted an integrated, collaborative approach to design. They offer new ways approaching design activity on a day-to-day basis. As part of an ICD approach, operations are geared toward exploiting opportunities for collaboration created by design chains (established using ICD principles, strategies and tactics), providing organisations with more effective ways of contributing to projects than traditional working relationships allow.

Practices

The methods used by design chain members to apply an ICD approach to their everyday activities. These include, depending on the time scale and scope of their influence, strategies, tactics and operations.

Principles

The core values that underpinning the integrated collaborative design approach. They provide a basic commonality of culture and attitude towards design management. Each of the three principles supporting the ICD approach must be established in a design chain member's culture before ICD strategies, tactics and operations can be successfully implemented.

Process

Continuous or regular actions taking place, or being carried out in a definitive manner, and leading to the accomplishment of some result; a continuous series of operations or repeating series of operations.

Project

The temporary environment formed by the collaboration of a number of organisations to achieve a common, defined end-result. When design chains are available, these organisations will, ideally, be members of a common chain, allowing their established collaborative practices to be applied to the project. Projects are usually concerned with physically modifying the environment, often resulting in the creation a new building or the alteration of an existing one. Projects exist in short, fixed periods of time, while the organisations undertaking them exist for the longer term.

Project Domain

The short-term environment in which design chain members collaborate to develop and deliver project solutions. By aligning their working practices and internal cultures within the business domain, the project domain provides the environment in which the advantages of their strategic relationships can be applied, letting the design chain achieve objectives that its individual members could not meet if working alone.

Provider

The provider produces design information for use by a receiver organisation. An organisation may perform a provider role at one level in a design chain while it may also perform a receiver role at another level in the same chain. It may also change its role depending on the current stage of the project. A provider/receiver relationship is generally expressed in a contractual relationship, but this may not always the case. Providers typically provide design information in response to design problems defined by others, although the provider may contribute to problem definition in certain circumstances.

Receiver

Within the context of the design chain, the receiver uses design information supplied to it by a provider organisation. Within the broader supply chain the provider would be supply goods or services, i.e. the supply and installation of building management systems. An organisation may be a provider in one tier of a design chain, while also performing a receiver role to another. Within the design chain, the good or service provided is design information. This produced in response to design problems, primarily defined by the receiver, although the providers may contribute to their definition in certain circumstances.

Resources

The assets (technical knowledge, production facilities, staff and financial capital for example) which an organisation owns and controls, combining them together to provide a product or service.

Role

Depending on whether an organisation, or organisation member, predominately generates or applies design information, the role of a provider or a receiver respectively will be adopted within the design chain.

Satisificing

A solution that satisfies the diverse needs of a set of problem stakeholders sufficiently to ensure they will all support the problem solution. A satisificing approach is one that recognises diversity in stakeholder expectations and requirements and therefore attempts to satisfy each stakeholder to a moderate, but not optimal extent. It is recognised that because of the probably diverse range of stakeholders, no single stakeholder is likely to be fully satisfied.

SCM Maturity Matrix

This is a graphical representation of the identified maturity levels of the key aspects of supply chain management. The matrix provides a framework for the allocation of work practices against the several key aspects required of supply network members according to the level of maturity they reflect.

Stakeholders

People or organisations with a project interest justified by their involvement in it or by its influence over them during construction.

Stakeholder Requirements

The functions that must be performed by the construction project output to ensure that it will satisfy the stakeholder needs. Project values are derived when stakeholder requirements are defined to be mutually acceptable.

Strategies

Those elements of an organisation's ICD approach concerned with planning the development of its ICD competency - by adopting successive ICD practices, while remaining responsive to the consequences of their adoption upon the development of their (and other design chain members') cultures.

Supply Chain

A project-specific grouping of organisations brought together to provide access to the competencies and resources required to deliver a project and satisfy its objectives. The organisations are sourced from a supply network. Further organisations may be engaged on a project where a supply network is unable to provide the required competencies.

Supply Chain Management

A collection of management practices underpinned by a philosophy that seeks to bring together resources drawn from a number of different organisations who together form supply networks and supply chains. The philosophy requires organisations to look beyond their organisational boundaries in order to optimise the overall delivery of a product or service to end-users/customers.

Supply Network

A group of organisations which have previously acted as provider to a receiver organisation that knows their competencies and to whom they are familiar. The supply network provides the framework and environment to govern business relationships.

Tactics

Tactics are used by an organisation to manage the implementation of its business development strategies while remaining responsive to short-term changes in its operating environments. These include the response of its internal culture and design chain partners to changes in its working practices.

Value

Value is achieved when needed functions are provided without unnecessary cost. It is sometimes expressed as the ratio of functionality to cost (value for money). Value management differentiates between stakeholder requirements that are needed to fulfil business need and those that are merely wanted. Because wanted requirements do not contribute to the satisfaction of business needs, they may be removed from the project to avoid the unnecessary cost that would otherwise be incurred by creating project elements to facilitate their performance.

Value, Business

The business benefit realised by an individual organisation within a design chain as a consequence of working within that design chain. These benefits relate the expectations of the business as a whole and may influence the work of all organisation members or just its management function, depending on the nature of the change or working environment that is causing the business value to be generated.

Value, Project

The project benefit realised by all project parties (by the procuring client /end user in particular) when the expectations and requirements of the project are satisfied through its delivery.

Value-adding Problem Solving Structure

The process constructed by a design engineer to complete a design task while responding to project values.

Value-adding Tool

A problem solving technique used to structure the consideration of the relationship between an emerging design solution and the project values it must satisfy. The use of these tools can be documented to create an audit-trail that establishes the role of individual design processes in providing value for money.

Value-adding Toolbox

A repository of value-adding tools maintained by an organisation in its business domain and made available to its designers (including those working in its collaborating design chain members) in its project domain. Designers use this resource to inform their use of value-adding tools to relate their technical design solutions to project values. The toolbox also disseminates examples of effective and ineffective value-adding tool use to facilitate corporate learning.

Value Chain

A generic model of the processes performed within an organisation to manage design and ensure that its specialist knowledge is incorporated into the design information passing through it. The model also represents processes associated with co-ordinating the design activity of design chain members, including managing the flow of design information across their interfaces.

Value Engineering (VE)

Within the ICD approach, the differences between value engineering and value management are considered insignificant, causing the two terms to be used synonymously. Value engineering is presented in this glossary because the term is in common use.

Value Framework

The collection of various value-related tools, techniques and principles of working adopted within ICD to structure the consideration of value within projects and to ensure that appropriate resources are maintained in the business domain so that they can be used in the project domain for the pursuit of value.

Value Management (VM)

The structured, usually facilitated, examination of the relationship of the overall project format to the client needs (representing his values) that must be satisfied. This process is typically administered through a series of workshops, allowing client representatives, relevant project stakeholders and key project team members to negotiate a satisfying project format. A variety of techniques is well established within the field of value management. Specialised individuals providing value management facilitation skills are often employed to manage the process and provide a source of these tools.

Value System

The strategic alignment of organisations to harmonise and align their business and project processes to optimise, for mutual benefit, the flow of design information passing between them.

Values

Within an ICD approach, values are defined as the needs (expressed as demands, objectives or requirements) of the project stakeholders. Values are shared among all stakeholders as a common understanding of the functional objectives they must achieve. This shared understanding is usually developed through stakeholder participation in discussion forums. Organisations and individuals also possess internal values that influence their working practices. Apart from the adoption of integral-value engineering principles as corporate values by the organisations (and organisational members) practising it, values of this type are not addressed by this specification because they do not relate to the value-adding toolbox.

Index

Index

designer, 3, 7, 10, 14, 19, 76, 89, 101, 105,
    110, 127, 129-130, 141, 146, 164, 167,
    193, 199, 206, 211
domain,
    business *see* business domain
    project *see* project domain
DSM *see* dependency structure matrix
DTI *see* Department of Trade and Industry

education, 49
Egan, J., 6
engineer, 7, 141
Engineering and Physical Sciences Research
Council, vi
end-user, 3-4, 33, 38-39, 51, 57, 60, 128, 157,
    183
EPSRC *see* Engineering and Physical Sciences
Research Council
European Foundation for Quality
Management, 40

facility visits, 137-139, 141
feedback, 70, 76, 81, 86-87, 89, 91, 107, 110-
    111, 127-128, 136, 145-148, 154, 163-
    165, 167, 170, 181, 188, 191, 198, 201,
    208, 213-214

gap analysis, 180

health and safety, 143, 147, 179, 180
human resources, 85

ICE *see* Institution of Civil Engineers
ICD *see* Integrated Collaborative Design
IDEF *see* Integrated Computer Aided
Manufacture Definition
information technology, 6, 16, 85, 96, 108,
    111, 146, 164
information flow, 3-4, 28-29, 30-31, 37, 41-
    43, 59-60, 65, 86, 89, 91, 125-126,
    130, 197, 199-200
Integral Value Engineering, 13, 16, 23-24, 28,
    35, 39, 43, 49, 55, 59-60, 81-86, 107-
    111, 114, 115-117, 119-120, 123-124,
    145, 147, 159, 161-165, 180, 181-183,
    205, 208, 210-213

Integrated Collaborative Design, vi-viii, 5-6,
    10-26, 33-35, 37-38, 40, 47-50, 57,
    58-60, 61-214
    adoption, 28, 35
    application, 45-55
    assessment, 40-43
    practices, viii, 18, 21, 23-26, 28, 37, 39, 44,
        45-55, 61-214
        operational level, viii, 5, 10, 18-19, 46,
            50, 52-55, 58, 59
        strategic level, viii, 5, 10, 18-19, 46,
            50, 52-55, 58-59
        tactical level, viii, 5, 10, 18-19, 46, 50,
            52-55, 58-59
    principles, viii, 5, 12-13, 18-19, 35, 40, 43,
        46-47, 50, 58, 83, 157, 198
Integrated Computer Aided Manufacture
Definition, 31, 130
interfaces, 4, 7, 25, 29, 31, 38, 41
interviews, 113-117, 139-141, 143
IVE *see* Integral Value Engineering
Institution of Civil Engineers, 142
Investors in People, 77

knowledge management, 49
KPI *see* Key Performance Indicator
Key Performance Indictor, 102, 139, 140, 142,
    143

legacy archive, 126, 128
Latham, M., Sir, iii, 6, 14, 200
Loughborough University, vi

management, 49, 76, 85, 101, 103, 105, 110,
    114, 125, 127-133, 135, 139, 141,
    149-154, 157, 173-175, 177-180, 200,
    203, 204
marketing, 49

network profile, 132, 136

Porter, M. E., 36, 37
percentage planned complete, 191, 192, 195
performance, monitoring, 26, 42, 49
PPC *see* percentage planned complete
price, 36, 48
    stability, 48

Index

design chains: a handbook for integrated collaborative design